Cavalry

Cavalry

ITS HISTORY AND TACTICS

Louis Edward Nolan

Introduction and Further Reading by Jon Coulston

WESTHOLME
Yardley

Westholme Publishing, LLC
904 Edgewood Road
Yardley, Pennsylvania 19067
Visit our Web site at www.westholmepublishing.com

ISBN: 978-1-59416-344-9
Also available as an eBook.

Printed in the United States of America.

TO

LIEUT.-COLONEL GEORGE WILLIAM KEY,

15TH HUSSARS,

THE FRIEND OF MANY YEARS,

THIS WORK IS INSCRIBED,

WITH AFFECTIONATE REGARD,

BY THE AUTHOR.

Contents

INTRODUCTION

Cavalry: its History and Tactics (henceforth simply *Cavalry*), first published in 1853, discusses not only the history and development of cavalry and mounted warfare, but also contains within it many elements of a program of reform proposed for Britain's horsed troops. It is an intelligent work, written by a widely read light cavalryman of exceptionally varied experience, which covers many aspects of equipment, training, drill, organization, formation and battlefield handling.

The author, Louis ("Lewis") Edward Nolan, would have been outstanding if only for his breadth of service.[1] Born in Canada in 1818, his early life was spent living in Edinburgh in Scotland, and Piacenza and Milano in Italy (then in the Habsburg Empire). He commenced his military career in 1832 as a cadet in the Austrian imperial army, training at the Engineers Corps School at Tulln, near Vienna. Commissioned into the 10th Hussars (Friedrich Wilhelm III King of Prussia's Regiment) which served in Poland and Hungary, he added Magyar to his already fluent French, German, and Italian. His position was secured by his father, ambitious for the prospects of his sons, three of whom served in the Austrian army. This was by no means unusual, for not only was the officer corps of the imperial army extraordinarily cosmopolitan, but it was said at the time that "the Hungarian hussar regiments have virtually become English colonies"![2]

In 1838 Nolan visited London for Queen Victoria's coronation and he was presented to the monarch by the Austrian ambassador. After

---

1. Career details are based on H. Moyse-Bartlett, *Nolan of Balaklava and His Influence on the British Cavalry*, Leo Cooper, London, 1971, and the excellent biographical Nolan website: www.silverwhistle.co.uk/crimea/nolan.html (consulted August 12, 2006).

2. Quoted in I. Deak, *The Lawful Revolution. Louis Kossuth and the Hungarians, 1848-1849*, Phoenix Press, London, 2001, 188.

witnessing the Hyde Park review, and considering his own family's service tradition, "young Nolan sought a more distinguished career in the British army."[3] After briefly serving as an ensign in the 4th Regiment of Foot, indeed illegally holding contemporaneous commissions in both the Austrian and British armies (April to October, 1839), he was gazetted cornet into the 15th King's Hussars in 1839. Purchasing his lieutenancy in 1841, he served in Britain, then continuously in India from 1843 to 1851, becoming regimental riding master in 1844. In 1848 he was elected to membership of the Army and Navy Club (London), established in 1837 by officers returning from India. He would have become familiar with its new Pall Mall building, opened in 1849, rather than with the previous St James's Square premises. Nolan's two books were at least completed at the Club. He purchased his captaincy in 1850, then returned to Britain having learned several Indian languages. On leave, he embarked on a tour of Europe (March-August, 1852) observing and participating in cavalry maneuvers in Sweden, Germany, and Russia. He was able to study Cossack and Circassian cavalry of the types which would later face the British army in the Crimea. Accompanying Nolan was his friend, Lieutenant-Colonel George William Key, to whom *Cavalry* was subsequently dedicated. On his return to Britain Nolan assumed command of the regimental base at Maidstone in Kent, and he led his regiment in the funeral procession of the Duke of Wellington in 1852.

The tour of European armies contributed to his new design for a service saddle (adopted by the British army after his death). His first book, *The Training of Cavalry Remount Horses: A New System* also appeared in 1852. Nolan's wide reading in many languages, his service in Europe and India, his research tour and his regimental experience, all went into *Cavalry*, published in the next year. This was an extraordinary statement on theory and practice of the mid-century period.

The work commences with a review of developments in cavalry types and tactics, with nods to the ancient and medieval worlds. It then

3. Obituary, *The Illustrated London News*, November 25th, 1854, 528. Quoted in full on a page linked to the website cited above.

goes on to the Early Modern period, looking at the continental use of cavalry during the Thirty Years War, under King Gustavus Adolphus of Sweden, and in the English Civil War. The latter was remarkable for the initial dominance of the horse-riding royalist aristocracy—great horsemen but undisciplined and difficult to control after charges— and the rise of professional, Parliamentary cavalry ("Ironsides") who eventually outmatched the "Cavaliers" in the field. Nolan joined British commentators who commonly praised Frederick the Great and his offensive use of cavalry in the Seven Years War, a king skillfully fighting off all the surrounding powers. However, Nolan is less usual in that he pays due attention to Austria and the wars with Ottoman Turkey on her eastern frontier. Indeed, Nolan so respected traditional Ottoman cavalry that he bemoans the influence of French training on nineteenth-century Turkish cavalry forces. At root here is a high regard for the mounted skills of Asiatic steppe nomads as they were practiced by Magyars, Turks, Tatars, and Cossacks right across eastern Europe and southern Russia. There follows a chapter which traces major cavalry actions through the Napoleonic Wars and makes reference also to British actions at Mudki (1845) and Chillianwallah (1849) in the Sikh Wars.

Succeeding chapters are thematic and deal with aspects of cavalry types, organization, riding styles, drill, formations and tactics, moving on to the functions of cavalry on the march, in camp, outposting, and skirmishing. The chapter on the charge, pursuit, and rally brings together much of the preceding discussion to concentrate on how cavalry may best be deployed offensively.

Apart from official manuals coming out of Horse Guards, such as *Regulations for the Instruction, Formations and Movements of the Cavalry* (London, 1833, 1844, 1851), there were very few cavalry treatises written in the first half of the nineteenth century by British authors. The genre of the military treatise is of course an ancient one. Works on battlefield arrays, orders of march and siege techniques were being written in the west from the 4th century BC at latest. Surviving texts on mounted warfare with schematic illustrations displaying ideal cavalry formations, in the manner of Nolan's *Cavalry*, date from the 6th century AD onwards.

Nolan was writing at the cusp of technological change in warfare, drawing upon the experiences of the Napoleonic Wars, Britain's wars in India, and continental suppression of the 1848 Revolutions. He also had an eye on firepower developments on the eve of the Crimean War, but it is clear from Chapter 14 that, like many of his contemporaries, he did not yet fully appreciate how radically the range and accuracy of the new weaponry would in future narrow the possibilities for traditional cavalry action on the battlefield. *Cavalry* formed a way-point in tactical discussions during the "long" nineteenth century, and fed into subsequent developments resulting from the Crimean experience, the American Civil War, the Austro-Prussian, and Franco-Prussian Wars. These discussions of the role of cavalry in battle were still not resolved as European armies entered the First World War.

Nolan might be compared with another Victorian British cavalryman, traveler, and writer, Captain Frederick Gustavus Burnaby. A renowned Guards horseman, Burnaby became a public name through his published accounts of mounted spy missions into Turkey and Central Asia. He died in action in the Sudan in 1885.

What adds particular drama and even pathos to *Cavalry* is Nolan's specific involvement in the Crimean War in 1854. Throughout the work there are statements and recommendations which speak directly to the coming war. When plans were being made to send a British army to support Turkey against Russia Nolan was a natural choice for someone to be sent on ahead to the eastern Mediterranean. Tasked with securing cavalry mounts, he toured Turkey, Lebanon, and Syria, and managed to assemble 300 horses and mules. He then joined the main force at Varna in Bulgaria and henceforward served as an aide-de-camp to the Quartermaster General, Lord Airey. He was present at the Battle of the Alma (20th September, 1854), and was the first man to be killed in the charge of the Light Cavalry Brigade at the Battle of Balaklava (25th October, 1854). Nolan, like Burnaby and so many other Victorian soldiers far from home, was buried in an unmarked grave.

DURING THE "LONG" NINETEENTH century, from the Napoleonic Wars to the First World War, cavalry experienced mixed fortunes in battle, and their functions were constantly debated and evolving. The Napoleonic Wars were predictably very influential in forming and sustaining tactical doctrine for decades. In Britain the Duke of Wellington was renowned for his low opinion of the performance of British cavalry in the Spanish Peninsular campaigns. At the great set-piece of Waterloo which sealed Napoleon's defeat and permanent exile, the British cavalry had initially performed well but were perceived as having been over-impetuous in opening themselves up to counter-charge and annihilation by, among other French mounted troops, Polish lancers. The need to husband slender cavalry reserves further acted as a limit on the independence and initiative of British cavalry commanders. This is particularly what Nolan railed against both in writings and in communications with his friend William Russell. Nolan and many other observers saw prompt decision-making and alacrity on the battlefield as all-important cavalry officer qualities. Cavalry must always be ready to exploit developing opportunities, must never stand to receive attack, and must always advance to meet an enemy building up momentum, even if the ground is broken and sloping uphill, just as it was when the Heavy Cavalry Brigade charged at Balaklava.

From at least the fifteenth century cavalry forces in European armies had developed in two main categories, "heavy" and "light," the former being aristocratic, heavily armored gendarmes (*sipahi* in the Turkish context), the latter light or unarmored scouts and skirmishers (Tatars and other steppe nomads in the East). As firearms grew more effective so armor declined in use, but the concept of close-order cavalry, large men on large horses, charging with heavy swords persisted into the eighteenth century. Helmets and cuirasses came and went in fashion, being retained longest in Poland and the Habsburg Empire where the horse archery of Turkish forces was still a formidable threat. The East also had a strong influence on European light cavalry developments, the Croats and Hungarians of the Habsburg frontier continuing to excel in mobile tactics, using lighter, curved sabres and retaining the flamboyant traditional dress derived from Asiatic fashions (pelisses, dolmens, sabre-taches, etc.). Into the Napoleonic period the two cavalry forms were still very much light and heavy. The former (cuirassiers and heavy dragoons) still often retained helmet and cuirass, and were expected to be held in reserve, en masse, until such time as an opportunity arose to charge shaken, disorganized, or retiring forces. Opposing formed cavalry were a prime target, but steady and prepared infantry, especially those who formed squares under cavalry threat, or artillery in prepared positions, were not likely to be swept away or overrun. The light cavalry (hussars, light dragoons, lancers) were intended still to picket, escort, scout, and skirmish, but were increasingly ready to charge other cavalry, even heavies in favorable circumstances.

For Nolan and other commentators, the important variables in the handling of cavalry were the size of man and horse, the weight of the man's equipment, the formation of the body of cavalry, the terrain over which it operated, and the alertness and alacrity of its commanders. Nolan was very specific on the latter element: "There is little time for thought, none for hesitation; and, once the movement is commenced, its successful accomplishment is the only thought allowed to pass through the mind of the commander." Heavy equipment could slow a heavy horse, and to be effective in the charge it was recommended that close formation troops use sharp, straight swords and the horse's momentum, but that they must be kept in formation and under close control. A disordered formation with slow horses was a recipe for disaster. The fluidity of mounted operations demanded that all forward formations should be backed by a reserve which could advance through them when they were tired, or on which they could retire if repulsed. Reserves were vital because without support cavalry would be swept away by fresh opposing forces. Clearly different tactics were necessary when cavalry faced different opponents, but enemy cavalry had to be faced offensively, never met at the stand, and infantry and artillery should be approached from a flank.

There was still much discussion after the Napoleonic wars, based on anecdotal experience and subsequent colonial battles, as to whether properly handled and determined cavalry could actually break steady infantry squares. Some cavalrymen advocated the lance as a balance-tipping weapon; others maintained that shear élan should see the cavalry through. In actual fact, the improvements in infantry firepower, alluded to by Nolan in *Cavalry*, had by the 1850s irrevocably shifted the balance. In one of the few occasions when British infantry in the Crimea might have been expected to form square in the face of oncoming Russian cavalry, the 93rd under Sir Colin Campbell at Balaklava bluffed an irresolute enemy into retreat with its "thin red streak tipped with a line of steel." Indeed, already in the 1840s there were clear indications that heavy and light cavalry were losing their traditional battlefield distinctions. As scouts and skirmishers the light cavalry had a future, but as a reserve to be unleashed in a shock-attack against other cavalry, disorganized or retreating infantry and artillery, the days of the heavy cavalry were decidedly numbered.

This is particularly clear in the developing designs of British cavalry swords which steadily came together for both heavy and light use. The Sword, Heavy Cavalry, Pattern 1821 had a slightly curved blade and a flat back, 36.5 inches long. It was double edged for the last 10.5 inches and had a steel bowl guard. The Sword, Light Cavalry, Pattern 1821 had a slightly curved blade and a flat back, 35.5 inches long. It was double edged for the last 10 inches and had a steel three-bar guard. Already the heavy, straight heavy cavalry sword and the curved light cavalry sabre were converging. However, for the first time, just before the Crimean War, the same sword began to be issued to all cavalry irrespective of regimental title. The Sword, Cavalry, Pattern 1853 had a slightly curved blade and a flat back, 35.5 inches long. It was double edged for last 9 inches and had a malleable wrought iron three-bar guard. For the Crimea the 11th Hussars and 2nd Dragoons were re-equipped at least partially, if not wholly, with the Pattern 1853. All other regiments in-theater acquired the new weapon swiftly after 1854, short supply being bypassed by Pattern 1853 swords being sent out from Britain having been taken off home regiments whose swords were replaced from old stocks of Pattern 1821. Subsequently, the three-bar guards were found to be brittle so the Pattern 1853 blade was combined with a bowl guard, larger than that of the Heavy Cavalry Pattern 1821, to make the Sword, Cavalry, Pattern 1864. Thus, the mix of sword patterns employed by the Heavy and Light Brigade regiments at Balaklava were all intended to be used in a heavy cavalry role, emphasizing use of the point rather than the cut. Nevertheless, there were complaints in the field of points not penetrating Russian clothing, or of blades bent when used to thrust. Overarching all of this was the remarkable business of metal sword-scabbards, remarked upon by Nolan in *Cavalry*. These scabbards blunted the blade inside and should have been replaced with wooden scabbards. Some regiments did not sharpen their swords between leaving Britain and going into combat in the Crimea!

The other offensive cavalry combat weapon, the lance, went in and out of military fashion in nineteenth century armies. It could be described as a hangover from the Turkish Wars of the Early Modern period, particularly in its survival in Poland and Russia. Cossack cavalry were lancers throughout, whether harrying Napoleon's retreat from Moscow, or scouting the allied forces in the Crimea. Polish and other light cavalry lancer regiments of the Napoleonic Wars impressed some British observers, not least by their activities at Waterloo where 15 percent of the French cavalry were lancers, so that the lance gained passionate advocates. There were widely publicized tests of the weapon and four British Light Dragoon regiments were armed with lances in 1816-17. It was also felt appropriate that lancers should wear Polish-style attire such as lancer trousers and the square-topped helmet (*csapska*). The belief was that the lance was a most effective weapon in the charge against opposing cavalry, that it enabled horsemen to spear crouching or prone infantry, and that it gave cavalry the extra reach necessary to lean in and disrupt an infantry square. Advocates argued about the optimum length of the shaft.

The lance was also a prominent weapon in some Indian military traditions, as observed by Nolan, so that lancer regiments, both British and native, were a natural part of the military establishments of India. It remained prominent in certain continental European armies, and, in the Americas, the Mexican army of the 1840s had at least a proportion of men in line cavalry regiments, and all those in companies of presidiales, lance-armed. One further regiment was dressed in full "Polish" attire, the Lanceros de Jalisco. This did not really impress the U.S. military establishment in the 1846-48 U.S.-Mexican War. There was only one experiment in lance-arming a cavalry regiment during the American

Civil War. Famously this was advocated by General George B. McClellan, perhaps enthused by his experiences as an observer in the Crimean War. The 6th Pennsylvania Cavalry (Rush's Lancers) fought one engagement in which it was routed and abandoned its lances. The regiment had a fine subsequent career, but as sword-and-carbine cavalry, not lancers. Although ill-trained, this regiment's debacle would seem to vindicate the complete opposition to use of the lance voiced by Nolan in *Cavalry*. Nevertheless, British lancer regiments reportedly performed well in India and Africa through the nineteenth century. The 17th Lancers was in the front line of regiments when the Light Brigade charged at Balaklava.

There was also some discussion as to whether the lance was a "light" cavalry weapon, as in Cossack use, or an impact weapon for use by charging, close formation heavy cavalry. The Austrians used Polish lancers with some effect against Hungarian hussars during the 1848 Revolutionary war. Russian cuirassier regiments had their front rank armed with lances for a heavy role. Also on the continent, the Prussians were so impressed with French lancers in the Franco-Prussian War that the lance was widely issued to German cavalry thereafter and into the First World War. Concerned by this development, the British army armed its cavalry as lancers for the western front. In Mesopotamia and Palestine the lance was carried mainly by Indian regiments, and by the opposing Turkish cavalry. The Yeomanry Regiments charged into battle with swords as their main weapon, so effectively terrifying the Turkish enemy that swords were re-issued to some regiments. Less convincingly, the Australian Light Horse rode into action waving bayonets!

The lance was always controversial, and many British cavalry in India and South Africa considering it a useless weapon for anything but the initial charge impact against other cavalry. Even then it was reported that, in the charge of the 16th Lancers at Aliwal (1846) during the First Sikh War, British lances failed to penetrate clothing. Moreover, the British also used a seven-foot weapon, unlike the continental nine-foot lance. Nolan, who suspected that lancers threw their weapon away in combat, is wholly dismissive in *Cavalry*: "All experiments with blunt lances on fresh horses go for nothing, in my opinion, for many of the thrusts would not go through a man's jacket; and in a campaign, when horses are fatigued, and will not answer the spur, even the skilful horseman is helpless with a lance in his hand."

Cavalry firearms were wholly unsatisfactory and almost derisory before the 1840s. Smooth-bore, flintlock, single-shot carbines, short-barreled for use on horseback, thus both short on range and accuracy, were difficult to load in the saddle, and a great encumbrance to the horseman in motion. They were, however, necessary for skirmishing with other cavalry and for picket duty. Single-shot, smooth-bore and even less accurate than carbines, pistols were favored in pairs for cavalry in the sixteenth and seventeenth centuries, but were regarded as retarding swift cavalry action in the eighteenth century. Into the nineteenth century lancers were equipped with them because their shafted-weapon supposedly precluded use of a long firearm. However, changing firearm technology did not only provide advantage to infantry and artillery on mid-nineteenth century battlefields. The development of multiple-shot revolving handguns, and revolving multiple-shot or breech-loading single-shot carbines, potentially revolutionized mounted firepower. These weapons had the effect of enhancing light cavalry and mounted infantry functions just when the traditional heavy cavalry role was being eclipsed, as a glance at some wars of the mid to late nineteenth century demonstrates.

For example, in the U.S.-Mexican War the Americans were initially very apprehensive of Mexican cavalry in terms of numbers, skills, and cultural traditions. On paper this was perfectly reasonable: there were nine Mexican line cavalry regiments, plus one regiment each of cuirassiers, hussars, and lancers, plus militia and presidial formations, as opposed to three regiments of U.S. Dragoons and various volunteer units. In practice the Mexican cavalry command was universally indecisive. Mounted forces were employed on the flanks but cautious leadership and opposing smooth-bore infantry firepower stopped cavalry attacks developing at Palo Alto and Buena Vista. Grand Napoleonic cavalry reserves were not committed either because the Americans held the initiative or the ground was too broken, as at Cerro Gordo and Chorubusco. In smaller scale encounters traditionally-armed Mexican cavalry were shocked by the firepower of opponents armed with revolvers, particularly those used by Texas Rangers and other irregular American cavalry.

The Hungarian Revolution had some superficial similarities to the American Civil War in that one side (Austria and Russia) eventually commanded overwhelming forces, while the other (Hungary) scrambled to patch together an army with limited resources, its antebellum army was unevenly split between the sides, and initially the best cavalry with a traditional socio-cultural advantage were with the insurgents. The hussar regiments were the cream of the Hungarian army, including Nolan's old regiment, the 10th Hussars, and, as he reports at the end of Chapter 4 of *Cavalry*, they could successfully take on the Austrian heavy cavalry. This is another facet of the traditional roles of heavy cavalry and light cavalry rapidly converging, here accelerated by the split between types of regiments on opposing sides of the conflict (not all the hussar regiments joined the Hungarian army but all the regiments in the Hungarian forces were hussars).[4] However, in most European armies by mid century the cavalry types were fulfilling the same close-order, heavy roles, in many cases to the detriment of the more mobile and flexible light mounted functions. The traditional titles still pertained, but were merely marked by the long-standing styles of uniform, no longer by field-exercise.

The American Civil War famously commenced with an unequal dichotomy in opposed mounted forces, as the aristocratic, rural, horse-riding and hunting cultural traditions of the south out-performed the much-parodied urbanites of the northern armies. However, the Federal cavalry trained hard and evolved its own tactical doctrines, taking full advantage of new technology, notably breech-loading, rifled carbines, but at the same time not completely forsaking the traditional cold steel of the *arme blanche*. Increasingly the Confederate cavalry faced better equipped, competent, and confident horsemen. Out-shooting southern

4. There is a huge canvas in the Budapest Military Museum (Hadtörténeti Intézet és Múzeum), Hungary, which depicts Austrian cuirassiers being pursued by Hungarian irregular cavalry and lassoed out of their saddles. The lasso is mentioned in *Cavalry*, and it was an Asiatic steppe herding tool applied to warfare in many periods.

cavalry and holding their own dismounted in broken ground against southern infantry, the Federal regiments even on occasion overthrew their mounted adversaries in traditional sword-charges. Flexibility of approach led to highly skilled, mixed-arms tactics whereby different companies of a regiment, or regiments of a brigade, operated mounted with swords or dismounted with carbines, in mutual support exploiting developing tactical circumstances. Indeed, towards the end of the conflict, when war-weariness was at its height, the Federal cavalry was most likely the only really innovative and motivated element in the armies of either side.

Nolan's writings were influential across the Atlantic in an America where the Civil War required that numerous mounted regiments be raised in a relatively short time, and that a doctrine be refined for their use. General McClellan's 1861 review of cavalry forces in Europe admired Austrian methods but made very little comment on British cavalry. However, although it recommended very few publications in English, Nolan's two books were among them. An American edition of *The Training of Cavalry Remount Horses* appeared in 1862. A general manual for the instruction of the newly raised U.S. Cavalry forces, written by J. Roemer and published in 1863, drew heavily on Nolan's writings, even down to its title. These related Northern publications in sequential years were followed in 1864 by a reprint edition of Nolan's *Cavalry* in the wartime Confederacy, published by Evans and Cogswell of Columbia, South Carolina. This company was an official printer of military forms and of Confederate currency, which virtually made the Nolan reprint a government publication for the edification of Southern officers.[5]

It is interesting to speculate what Nolan would have made of the Cavalry of the American Civil War had he survived Balaklava. He surely would have made the comparison between the shift in balance of skills

5. G.B. McClellan, *European Cavalry, Including Details of the Organisation of the Cavalry Service Among the Principal Nations of Europe*, J.B. Lippincott & Co., Philadelphia, 1861; K. Garrard, *Nolan's System for Training Cavalry Horses*, n.p., New York, 1862; J. Roemer, *Cavalry: Its History, Management and Uses in War*, Van Nostrand, New York, 1863; L.E. Nolan, *Cavalry: Its History and Tactics*, reprint from the 2nd London edition, Evans and Cogswell, Columbia, SC, 1864.

and victory in the Union's favor with the rise of the Parliamentarian cavalry in the English Civil War. He would also have applauded the simplification of uniforms and equipment, and the hard-riding, long-distance endurance of mounted troops operating independently across country. It almost goes without saying that he would have approved of the continued use of the sword and the discarding of the lance. In the 1860s British officers were inclined to lionize the Confederate commanders; their struggle against great odds and their supposed gentlemanly ethos appealed to men who themselves could only hope to command armies small by European standards.[6] Judging by *Cavalry*, Nolan would have provided a shrewd and even-handed treatment of mounted warfare developments. He would have admired the successful operations of James Ewell Brown Stuart, but have deprecated the mistakes before Gettysburg. He would probably have cited the activities of John Singleton Mosby and Nathan Bedford Forrest as good examples of light cavalry potential. Undoubtedly he would have analyzed the events around Brandy Station with relish.

Cavalry on both sides in the Austro-Prussian War of 1866 were ineptly deployed and committed. Austrian light cavalry were partially effective in masking their North Army in the face of the dispersed Prussian armies which crossed into Silesia-Moravia. However, at the decisive battle of Königgrätz the Austrian heavy cavalry reserve was routed in thirty minutes by artillery and infantry firepower. The Prussian reserve cavalry which had the opportunity of fulfilling its proper battlefield function of being unleashed on a retreating enemy was held up by incompetence and traffic congestion, so out of 350 Prussian squadrons only 39 clashed with the Austrian cavalry, and, after the latter's flight, there was no further attempt at pursuit. The half hour of hard charging ended the dispute within the imperial establishment between the traditionalists who favored mass cavalry charges and the innovators who advocated the use of small light cavalry formations armed with modern carbines, in the latter's clear favor.

6. I.F.W. Beckett, *The Victorians at War*, Hambledon and London, London and New York, 2003, 3-4.

By 1870 the Prussians had actively applied the lessons of this conflict. Their cavalry were deployed forward in reconnaissance divisions keeping close to the enemy, their officers well trained in tactics, scouting, languages, and in exercising their initiative. Few regiments were retained for heavy service. The French on the other hand still maintained the old cavalry doctrine and ethos with disastrous results. At Froeschwiller a heavy brigade charged Prussian infantry who did not even form squares, but fired in line and incapacitated 800 out of the 1,200 cuirassiers, none of whom made it to within fifty yards of the infantry. At Mars-la-Tour a Prussian cavalry brigade charged French corps artillery and overran it by making subtle use of undulating terrain on the approach. The action was critically successful, but more than half of the 800 horsemen fell in the charge. Lastly, at Sedan, a French light cavalry brigade was flung forward in an attempt to break through encircling Prussian forces. Three charges were made against 144 guns and infantry in line with the result of utter carnage among the French.

In the next great European war cavalry seldom had the chance to scout or act as massed, break-through forces in the western theater once the trench systems had been established from Switzerland to the North Sea. On the odd occasions that they were deployed casualty rates inflicted by twentieth century artillery and machine-guns were predictably high. Many cavalry regiments were dismounted for use as infantry in the trenches. However, in the eastern theatres there was still scope for more traditional activities and by this time cavalry formations also had their own integral machine-guns for support. In Mesopotamia the cavalry were employed to outflank Turkish positions during British advances up the Tigris, but it was in Palestine in 1917-18 that the mounted arm had its swan-song. The 4th Australian Light Horse Brigade charged and overran entrenched Turkish infantry in front of Beersheba, a critical part of General Allenby's brilliant victory in the Third Battle of Gaza. The Australians were essentially mounted infantry which dismounted to clear trenches of their remaining defenders. However, in a small but glorious action at Huj, ten troops of Warwickshire and Worcestershire Yeomanry charged Austrian and German artillery supported by machine-guns and Turkish infantry. They advanced using the ground to partially mask their approach, nevertheless, as at Mars-la-Tour, losses were massive. Some 100-140 out of 170 horses involved were killed Nevertheless, 120 sword-armed cavalry succeeded in capturing eleven artillery-pieces, four machine-guns and 70 prisoners.

These latter actions were exceptional and the history of cavalry in the nineteenth century is one of evolving functions in the face of increasingly efficient infantry and artillery firepower. Extended ranges of firearms in particular transformed the situation so that the largest targets, the horses, were brought down with fearful losses long before charging cavalry could come into contact. This resulted in greater emphasis being placed on the traditional light cavalry functions, and on horsed troops also taking advantage of firearm technology to act as fast-moving mounted infantry. Indeed there was a short period around the American Civil War when their breech-loading carbines gave them local advantages over infantry with muzzle-loaded rifles, and a real capacity to take and hold ground against infantry counter-attack. In the colonial wars of the later nineteenth century, for example in Afghanistan, South Africa, and on the North American plains, the invaluable mobility of mounted troops even in broken terrain was fully recognized.

THE BRITISH ARMY WHICH FOUND itself having to mount an expedition to the Black Sea region in 1854 was, and may still be, easily criticized, even parodied, as aristocratic, dilettante, complacent, incompetent, outmoded, and unprepared. This is the line taken splendidly by Tony Richardson's film, *The Charge of the Light Brigade* (GB, 1968), which did for the Crimean War what *Oh What a Lovely War* (GB, 1969) tried to do, musically, for the First World War.[7] In the words of H.D. Arnold-Forster the army was "a social institution prepared for any emergency except that of war."[8] This was an army officered by the system of purchase of commissions and promotions. Although not necessarily all bad (having, it

7. Characteristic is a quote put into the mouth of senior command: "It will be a sad day for England when her armies are officered by men who know too well what they are doing- it smacks of murder."

8. Quoted in B. Farwell, *Mr Kipling's Army*, W.W. Norton and Company, New York and London, 1981, 11.

must be admitted, served the army well by supplying the likes of Wolfe, Wellington, and Nolan!), purchase could lead to a geriatricization of senior command, and to the holding back of talented, poorer officers. Already in 1856 a Royal Commission report condemned it as "vicious in principle, repugnant to the public sentiment of the present day, equally inconsistent with the honor of the military profession and the policy of the British Empire, and irreconcilable with justice."[9]

By mid century that public interest in the army had grown enormously, fanned by the development of an industrial, urban middle class, which read newspapers avidly. Curiosity was heightened further by a series of high-profile scandals concerning the conduct of aristocratic officers within regiments, most notorious of whom was James Thomas Brudenell, 7th Earl of Cardigan. Added to this was a super-power war covered by newspaper correspondents in an unprecedented level of detail and short communication time from battlefront to breakfast table, in which key reporters such as William Howard Russell (1820-1907) of *The (London) Times* were unsparingly critical of the aristocratic army command. Thus public attention to such small details as the transmission of orders at the Battle of Balaklava eclipsed any equivalent interest in the Napoleonic Wars (for example, circulation of *The Times* in the early nineteenth century was 5,000; at mid century it was 40,000). Now when there were reverses and failures the public could be whipped up by the press into a witch-hunt. The Crimean War was already signal for the press-led public demand for war against Russia in support of poor, bullied Turkey. With woefully inadequate logistical systems and medical infrastructure, as revealed especially by the disastrous winter of 1854-55, the Crimean War became, in the British public mind, both the nadir of army incompetence and the proof of the need for radical army reform. Brave sacrifice by common soldiers ordered into disaster by aristocratic duffers seemed to be epitomized by the Charge of the Light Brigade.

The beneficial legacy of this public concern was the call for reform and a major shift in the perception of British soldiers, a shift, to use the terms of Rudyard Kipling (1865-1936), from 'beggars in red' to 'gen-

9. Farwell, *Mr Kipling's Army*, 57.

tlemen in khaki.' Of course army reforms did not guarantee improvements in high command efficiency and initiative, as the Anglo-Boer War would demonstrate, nor did they necessarily provide state-of-the-art medical infrastructures, as the Mesopotamian campaign of the First World War was to prove. .

Not only newspapers but also photographs helped to concentrate public awareness. The Crimean War was not the first conflict to be recorded in the nascent art of photography, the 1848-49 European Revolutionary Wars and the 1846-48 U.S.-Mexican War preceded, but it was the first to be photographed in detail. Plates made by Roger Fenton (1819-69) and others recorded landscapes and base-camps, whole battalions mustered, and formal studies of individual men wearing both regimentals and the more improvised clothing of a harsh campaign. Thus they recorded some of the winter hardships, and the optimistic camaraderie of brother officers or soldiers grouped around fires and formation mascots, soldiers stiff in the long-held poses demanded by contemporary photographic technology, yet radiating professional pride and good humor.

There were of course elements of truth in the parodies, but they were neither fair nor balanced, and, counter-intuitively, the British Army was actually one of the most technologically up to date and innovative forces on the planet. Although the game played by modern military historians of championing a specific conflict as the "first modern war" is rather anachronistic and unhelpful, there is a cogent argument in support of the Crimean War for this "title" (which is usually awarded to the American Civil War by America commentators). The industrial development of Britain and France was reflected in the employment of rifled steel artillery, rifled muskets, steam ships, ironclad vessels, telegraph communications, and (eventual) use of railway transport. The fieldworks, scale, and longevity of the Sebastopol siege made it more like the trench warfare of the First World War (and some American Civil War sieges) than formal eighteenth and early nineteenth century sieges. Considering the shortness of the war the numbers of troops mobilized (two million Russians for all fronts) and the number who died (up to 640,000) were enormous.

In contrast to the paucity of British cavalry literature between the Napoleonic and Crimean Wars, published manuals and service journals fairly buzzed with discussions of firearm technology regarding both the infantry and the artillery arms. Research and development have been characterized succinctly by Huw Strachan: "The interaction between experience in the Napoleonic Wars, subsequent European theory (which was itself built on that experience), and imperial practice helped shape the tactics of the British army. But towards the end of the period the crucial determinant of tactical development became technological change, which effected a revolution in the power of small arms and promised another in that of artillery. These innovations were, in the parlance of economic historians, demand-driven. The army itself generated the enthusiasm for rifled firearms, although the doctrinal consequences of the introduction remained confused. Arguably, in allowing technology to determine tactical doctrine, rather than doctrine to create its own tactical demands, the pre-Crimean army set the pattern for its successors up until our own day. Certainly it was not primarily the Crimean War which initiated tactical change."[10]

Thus the majority of British infantry regiments in the Crimea carried the Pattern 1851 Minié rifled musket, which had both increased penetration and increased accuracy over the traditional smooth-bore Brown Bess musket. The latter could score 66-70% hits at 100 yards but only 25-30% at 200 yards, even under ideal firing-range conditions. The Minié rifled musket was effective up to 1,600 yards (according to the Russians who suffered its fire) and eleven times more accurate than smooth-bores at 400 yards. It has been calculated that at the Battle of Vittoria (1813) in the Spanish Peninsular War one bullet in 459 at best hit its target, while at the Alma (1854) one in sixteen was effective. Russian infantry regiments in dense columns (armed with smooth-bores) were swept away at the Battles of the Alma and Inkerman. Such weapons were subsequently to exact such a high price from infantry and cavalry formations in the American Civil War.

10. H. Strachan, *From Waterloo to Balaclava: Tactics, Technology, and the British Army, 1815-1854*, Cambridge University Press, Cambridge and New York, 1985, ix.

The British expeditionary force to the Crimea consequently exhibited variable qualities and defects, drawn as it was directly from the home country. Quite understandably it lacked much in the way of strategic planning, staff-work, logistical and medical support, and field experience. Nothing like this scale of amphibious operation had been attempted since the Peninsular War and the War of 1812.[11] Moreover, the methods of grand tactical training familiar to modern armies were not yet standard in Britain. The co-ordination of formations working together was practiced more through reviews and state ceremonial than through field maneuvers. The principal, and, as it turned out, most timely, exceptions were the two camps held at Chobham Ridges, southwest of London, in June and late July 1853. The first, and largest, involved one cavalry and three infantry brigades, 8,000 men, exercising over five weeks, and included a number of cavalry regiments which later served in the Crimea (2nd and 6th Dragoons, 4th and 13th Light Dragoons, 8th Hussars, 17th Lancers).

The inexperience of command, and of rank and file in the field, was demonstrated after the Crimean War had commenced. Once the British expeditionary force had reached Varna in Bulgaria, Lord Cardigan led a cavalry *razzia* in force up to the Danube and back. Covering 300 miles in seventeen days, the "Soreback Reconnaissance" was disastrous to the health of men and horses involved because they were pushed too hard and the return route was circuitous. In contrast, similar marches were efficiently carried out by fit and experienced troops in Indian climatic conditions. In the Crimea itself observers were quick to contrast the field-craft of British soldiers unfavorably with that of the French. This was principally because the latter were fresh from service in Algeria, and their army had a Napoleonic ethos of "living off the land" (not an easy prospect in the Crimean countryside). However, when British organization did improve into 1855, then it was the French who fell behind in supply and camp organization. In the field the British cavalry

11. See P. Hore (ed.), *Seapower Ashore. 200 Years of Royal Navy Operations on Land*, Chatham Publishing, London, 2002; C.D. Hall, *Wellington's Navy. Sea Power and the Peninsular War, 1807-1814*, Stackpole Books, Mechanicsburg, 2004.

were sometimes found wanting in basic field procedures. The advance after the Battle of Alma, passing round Sebastopol, was a confused mess of uncoordinated columns and poor scouting during which the Commander-in-chief, Lord Raglan, was almost captured by the Russians. During the Battle of Balaklava the Heavy Cavalry Brigade advanced without scouts or flankers, and thus was surprised by a large body of Russian cavalry approaching over a ridge.

However, it may be claimed, with qualifications, that in the three battles fought by the British on the Alma, at Balaklava, and at Inkerman, the army performed admirably by contemporary standards. Of the three, only Balaklava gave the cavalry opportunities for action. Inkerman was entirely an infantry struggle. At the Alma the Light Cavalry Brigade took an inland, left flank position, but was not committed to pursue the defeated and fleeing Russian army towards the end of the day. Understandably, the British joint commander, Lord Raglan, was much criticized in British cavalry circles for this "failure." In Raglan's defense it might be remarked that at this stage of the campaign the brigaded British light cavalry regiments were the only mounted troops in the entire Anglo-French army (the British Heavy Cavalry Brigade was landed at Balaklava later), and that to unleash them over the broken and steep hills to the front would have been a serious risk. Raglan would also have been aware of the low regard with which his mentor, Wellington, had held the light cavalry, viewing them as "mere brainless gallopers." In the Spanish Peninsular, and especially at Waterloo in 1815, the British cavalry had developed a reputation for uncontrolled abandon in the charge, and an inability to rally early enough to avoid disaster from enemy counter-attack.

At Balaklava the Heavy Cavalry Brigade, less than 600 strong, charged, outflanked and cut through a superior mass of 1,700 Russian dragoons and Cossacks. The latter were cautiously led and the British enjoyed complete ascendancy despite unfavorable ground. Losses were actually slight on both sides, but the moral effect of defeat on the Russians was decisive.

Later in the day the Light Cavalry Brigade of five regiments (4th and 13th Light Dragoons, 8th and 11th Hussars, 17th Lancers) was misdirected forward, not east-south-east up to redoubts on the Causeway Heights in order to recover captured artillery, but east down the floor of the North Valley. Here they faced prepared artillery directly supported by a mass of cavalry, with artillery, infantry, and more cavalry on the north and south flanking slopes. The advancing British troops were operating in close order under strict control, one regiment equipped with lances, at least some of the others with the hybrid light-heavy Cavalry Sword, Pattern 1853. As they accelerated they were not actually an easy target for guns to the front, and even less so moving across the sights of guns on the flanks. Infantry smoothbore muskets did not command the full width of the valley. The brigade successfully reached the guns and eliminated the crews, then cut their way through the supporting cavalry. Unsupported, elements of the Light Brigade then returned back down the valley but further defeated a body of Russian cavalry which had moved down off the flanking ridges and formed up in a position now in front of the retiring British force. These too were cut through and the brigade returned to its start-point.

Of the 658 men who commenced the exercise, 287 were lost, fewer than was first reported, as unhorsed and less severely wounded men trickled in. Less than 113 horses survived, again fewer than in early reports because many wounded animals had to be destroyed subsequently. Traditionally this has been seen as a tragic blunder, throwing away the flower of the light cavalry. More dispassionately it may be judged to have been a signal tactical success, vindicating the training, discipline and cohesion of the British regiments and confirming their complete ascendancy over Russian cavalry. The overall contribution of the charge may have been to reinforce the pusillanimity of Russian command and to end for the rest of the year any renewal of credible threat to the British Balaklava naval base. The Light Brigade was fortunate not to have been facing an enemy equipped with more up-to-date weaponry. Its advance and extraction were not entirely unsupported, for a charge by French *chasseurs d'Afrique* onto the Fedioukine Heights neutralized Russian artillery on the brigade's north (left) flank. Had the Cavalry Division commander, Lord Lucan, followed his orders to the letter, then the Heavy Cavalry Brigade, with attendant horse artillery, would have fully supported the Light Brigade, just as Nolan would have recommended. The Russian forces in the valley would have completely routed, although those on the ridges would probably have been safe and able to inflict casualties throughout. Would this ultimately have been a further waste, throwing the rest of the cavalry away? In a sense this is and was an entirely academic question. The cavalry were not called upon for the rest of the year. However frightful the loss of horses in the Light Brigade, and in a putatively committed Heavy Brigade, almost all the mounts were to die through exposure, neglect, and starvation through the following winter.

Returning to Nolan, this officer, advancing with the Light Brigade, was the first to fall to enemy fire, and he thereafter rode into a history of controversy. By the 25th of October, Nolan, a spectator of the Battle of the Alma, was a man wound up by frustration at what he perceived as the sluggishness, lack of enterprise, and the ability to lose opportunities for decisive action displayed by the British high command, from Raglan downwards. So Nolan, perhaps the most intelligent, experienced, and professional cavalryman in the whole army, carried Raglan and Airey's order, from a C-in-C whose view was well known (and specifically derided by Nolan to William Russell) that the cavalry should be "kept in a band box," i.e. safely back and uncommitted. He rode to Lord Lucan, who was known derisively for his caution as "Lord Look-On." Lucan and his immediate subordinate, the Light Brigade commander, the Earl of Cardigan, hated each others' guts. They were brothers-in-law. They were quite incapable of coolly debating the exact meaning of the order, once the direst interpretation of it had been erroneously made. Nolan, strident with exasperation at what he saw as vacillation, became intemperate, and did not function properly as an aide-de-camp. He gesticulated wildly, failing in his duty to verbally explain the commander's intentions as an accompaniment to the written word. Frustration and seething contempt perhaps combined in Nolan with the hope that at last the light cavalry would be given its head. However, the suggestion which has been made that Nolan deliberately misdirected the charge away from the causeway redoubts towards the Russian artillery and cavalry at the far end of the North Valley, out of an academic conviction that resolutely led cavalry could sweep all before them, is quite unsustainable.

Nolan was a man who venerated British cavalry service and preached the kindness regime of horse-training. His fault was to fail to adequately explain the commander's intentions which were formulated from an all-seeing rearward position on the Sapouine Heights. The only reasonable interpretation of Nolan's behavior once the brigade was committed was that he belatedly realized what was unfolding and acted immediately to correct the misdirection. With good reason bugles were used to sound cavalry orders with notes which cut through the thunder of hooves and the metallic jingling of harness and scabbards. Thus it will never be known exactly what Nolan was shouting through the noise as he veered across the brigade front, and thus fatally into the path of shrapnel from the first Russian shell to burst.

Attempts seem to have been made to make Nolan a scapegoat in the immediate aftermath of the battle and something of this stuck, as may be seen in the diary of Elizabeth Grant, a society lady living in Dublin. Her entry for 15th November, 1854, three weeks after the battle, comments that "Lord Raglan bade General Airey, the Quarter Master General, to send to the Light Cavalry Brigade to advance position and remain on the defensive. General Airey sent the message by his Aide de Camp, Captain Nolan, a very dashing Cavalry Officer, whose peculiar monomania it was, poor man, to imagine the horse invincible. He either mistook or misinterpreted his message, for he desired Lord Lucan to advance and charge....One hundred and sixty Light Horse returned out of that doomed band; a hundred came in afterwards. Only a hundred and thirty were killed outright; the rest were only wounded. Thirteen officers fell; the first among them was the authour (*sic*) of the butchery, Captain Nolan. The correct order was found in his pocket".[12]

However, friends and colleagues swiftly rallied to Nolan's defense, and already in obituaries quotes from *Cavalry* were employed to demonstrate the absurdity of charges of wilful misdirection. In particular the assertions that you should "never attack without keeping part of your strength in reserve" and that "charges on a large scale should seldom be

12. P. Pelly and A. Tod (ed.), *Elizabeth Grant of Rothiemurchus: The Highland Lady in Dublin, 1851-1856*, New Island, Dublin, 2005, 239-40.

attempted against masses of all arms, unless they have previously been shaken by fire" were singularly trenchant. The last word should be left to an anonymous officer, written three days, not three weeks, after the battle, and quoted in *The Illustrated London News* obituary: "Poor Lewis Nolan has gone to his rest. In a cavalry action three days ago he bore an order from Lord Raglan to Lord Lucan to charge a battery of heavy field-pieces, and in the act of delivering it, a piece of shell struck him on the left breast, and passed through his body. Death, by the mercy of Heaven, was instantaneous. Poor Lewis! He was a gallant soul. The day before his death, I am glad to think, I met him, and he said, 'Well, Bob, is not this fun? I think it is the most glorious life a man could lead.' Few men of his years promised to be such an ornament to his profession. I am sorry to say, now that he is gone, some people here say that in the heat of the moment, poor Lewis gave Lord Lucan a wrong order. Such is not the case. The order was a written one, and therefore the mistake was not on his side."[13]

Nolan's short life thus ended at a pivotal point in history, of tremendous interest to students of detailed cause and effect, but it is most fortunate that the man left to posterity much more than a cloak lent to William Russell and the penciled scribblings of a campaign journal. Nolan's writings on cavalry were the product of long and deep reflection by a consummate horseman who loved horses and profoundly respected the military service. *Cavalry* stands as a classic of its genre, and, despite its backwards look at the glories of mounted warfare, it joins so much other evidence demonstrating that the officers of the mid-nineteenth century British army were not the buffoons of popular fiction but professionals dedicated to innovation and improving their advantage over any adversary.

Jon Coulston

13. *The Illustrated London News*, November 25th, 1854, 528.

PREFACE

THERE IS, PERHAPS, NO BRANCH of the whole science of war which has engaged so little of the attention of military writers as that which refers to the formation and employment of cavalry; while at the same time it must be admitted, that few branches of the service are really more important than this.

On looking, however, closely at the subject, it is easy to understand why this should be the case. The tactics of cavalry are not capable of being reduced to rule, like the mechanical operations of the engineer, or even the slower and more methodical movements of infantry.

With the cavalry officer almost everything depends on the clearness of his *coup-d'oeil*, and the felicity with which he seizes the happy moment of action, and, when once action is determined upon, the rapidity with which his intentions are carried into effect. There is little time for thought, none for hesitation; and, once the movement is commenced, its successful accomplishment is the only thought allowed to pass through the mind of the commander.

Much, then, must partake more of the inspiration of genius than of the result of calculation and rule. Still there is a great deal in the profession of a cavalry officer which can be reduced to writing, and which it is most important he should know and carefully study; many things which his knowing beforehand will enable him to profit by, or, having carefully thought over, will suggest to him, in the hour of need, expedients that no amount of service, or of thought, would ever enable him to perceive without previous reading and long study of the subject.

Nothing, however, can well be more difficult than to attain to this in the present state of the literature on the subject. Few, if any, special books exist, as before stated, treating exclusively of cavalry, and none certainly of any importance in the English language, so that the student must pick out for himself, from the histories of campaigns and battles, or from general treatises on the art of war, those parts he stands in need of; and as these are seldom written by persons intimately acquainted with either the difficulties or advantages of this arm of the service, he must elaborate his conclusions for himself, and often from the most imperfect and erroneous data.

It is not pretended that the present little treatise will supply this desideratum in military literature, or serve as anything but a stop-gap to supply for a time the place of some more worthy treatise, which, it is hoped, may before long occupy its place. Still the author, having served in the continental cavalry, and with our own in India, and having thought much on the subject during a tolerably extended acquaintance with the cavalry of various nations, hopes that he may not be deemed forward in contributing his mite towards an improvement in the literature of Cavalry, and in offering such suggestions as he hopes may assist in bringing forward this important arm to the level of the intelligence of the age, and to the improved condition to which all branches of the service must be brought, if they are to compete successfully with their rivals in the next great struggle that may take place.

The sudden transition from peace to war is a critical moment for all armies, but more particularly to those whose officers are deficient in the theory of their profession.

Take for instance our cavalry in the last great European war; they were superior to that of most nations in the headlong courage of the men, the quality of their horses and equipments, but unfortunately inferior in tactics: the published despatches of our greatest commander bear too frequent testimony to the fact that our officers often neglected to provide reserves when they charged, or to take other necessary precautions, the want of which entailed occasional defeat upon our troops, in spite of the determined bravery which they displayed upon all occasions.

A deficiency in the theory of war entails the necessity of gaining experience at a heavy loss of life when war actually breaks out, and it is, therefore, the duty of every officer to endeavour to gain a knowledge of his profession before he is called upon to take part in that game where every false move is attended with more or less disastrous consequences to his country, and to the soldiers under his command.

In this work I have endeavoured to gather knowledge from the recorded experience of others, and to collect from various authors on cavalry warfare much that may assist young officers in gaining a knowledge of their profession. At the same time, I hope it may be understood that the introductory sketch of the history of cavalry is not meant to supply the place of what would form one of the most brilliant and interesting works on the history of the art of war; but is merely intended to suffice, by stating what has been done, to make it intelligible—what it is thought may again be accomplished, when better understood tactics, and better equipments, have restored the horseman to his proper position in European armies.

The remarks on dress, equipment, drill, etc., are merely the application of the principle of common sense to objects which, though useful, perhaps indispensable, when first introduced, have become positively noxious in the course of time, when they have long outgrown the purpose for which they were introduced, but are still allowed to linger because change in what is so thoroughly organised is always difficult, and sometimes dangerous; but certainly neither so difficult nor so dangerous as a blind adherence to exploded theories or antiquated usages, which must eventually be abandoned on the first rude shock of war.

On the tactics of the cavalry it is difficult to say much that is either very definite or satisfactory; but the use of the arm is generally so little understood that it is important (even to attempt) to place in a clearer light what is known and admitted on the subject than has hitherto been done; and, although the author cannot flatter himself that he has produced much to effect this, he hopes he has made a step in the right direction, and pointed out a path which will (as in all other subjects) be of suggestive interest to many of his brother officers, and which may be followed with more effect by others who have more leisure and more of those peculiar qualifications than he can pretend to for successfully handling so important and difficult a subject.

The work was written as leisure occurred in the intervals of duty and of other avocations, and, as might be supposed in a book so composed by a person little acquainted with the art of bookmaking, was somewhat disjointed in its arrangements.

I am indebted to Mr. James Fergusson for the present mode of arrangement, as also for sundry hints by which I have profited in putting the work together; and for his candid criticism and friendly assistance I shall ever feel grateful.

I also thankfully acknowledge the valuable assistance I have received from Mr. Charles Mac Farlane, whose love for the service is well known. He corrected many errors in the MS., and saw the book safely through the press.

I have conscientiously drawn out whatever passed under my observation, and in writing of the different systems prevalent in the cavalry I have stated freely, and without reserve, what I believe to be the truth; this, I trust, will not give offence, or hurt the feelings of any individual; and if my remarks prove useful to the service to which I have the honour to belong, then my object in writing this book will have been fully obtained.

Louis Edward Nolan
Army and Navy Club, 1st July, 1853.

1

Historical Sketch

Great Cavalry Actions of Ancient and Modern Times

T HE HORSE, INDIGENOUS TO THE COUNTRIES of the East, attained at
a very early period to that development, size, strength, and beauty
of form, which he did not reach in any other regions until after the
course of many centuries, and the application of numerous artificial
means, and an uncommon and persevering attention. It should appear
that in every part of Europe the horse was but a ragged pony, or, at
most, a rough, stinted galloway, when the horse in Arabia, Asia Minor,
the vast plains of Mesopotamia and Persia, was a splendid animal, well
suited to the purposes of war, and much used in battle. All these Eastern
nations were, in fact, equestrians, and made nearly an exclusive use of
cavalry in their wars with one another.

The horses of the East were brought into Europe by the Greeks and
their neighbours, across the narrow straits of the Bosphorus and
Hellespont, or were carried across the Ægean Sea, from the Greek
colonies in Asia Minor. The breed was propagated extensively in Thrace,
Thessaly, Macedonia, and other regions. The Athenians and most of the
Greeks imbibed a passion for beautiful horses and horse-racing; and
more than a century before the time of Alexander the Great the Greek
sculptors must have been familiarised with the forms and action of high-
bred horses, or they could not have left behind them the noble Elgin
marbles. But it does not appear that the Greeks at this period made any
extensive use of cavalry on the actual field of battle.

The Romans subdued their neighbours by infantry. In Livy's account
of their wars with the Samnites and other peoples, we read of cavalry,
and war-chariots drawn at the gallop by horses; but it should seem that
the horses were few in number and small in size, that their war-chariots

were as primitively rude and as ineffectual as those of our painted ancestors in Britain, and that such cavalry never really won a battle or decided
a great victory.

In short, cavalry was little used by the Greeks and Romans until war
brought them respectively in contact with Asia and the contiguous
regions of Africa; and it was in Alexander's campaigns, and the Punic
wars, that these nations first learned, from their well-mounted enemies, the value of good horsemen.

The first formation of the Greeks and Romans appears to have been
the same: they formed their horsemen in oblong squares, or in the shape
of a wedge, with the idea that the leading file, carried forward by the
mass in their rear, would break through all opposition. This system of
charging in a wedge has been since adopted among the Turks and other
nations distinguished by their equestrian qualities.

When the great Alexander passed over into Asia for the conquest of
the far-spreading Persian empire, his army was composed chiefly of
infantry, formed into phalanges. The Macedonian phalanx will be for
ever memorable in military history; but there is no distinctive
Macedonian formation of cavalry to share fame with it. Once in Asia
Minor, a country of fine horses, and among the Greek colonies thickly
scattered from the mouth of the Hellespont to that of the Granicus, and
from the mouth of that river onward to the confines of Syria, it must
have been easy for the young conqueror to augment his cavalry. No
doubt, between his first landing in Asia and his first battle, that arm was
both increased and improved.

At the battle of the Granicus he made 4500 or 5000 horse ford the
river. The Persians on the opposite Lank opposed them bravely, and
fought hand to hand, but were at last driven off the field, chiefly, it is
said, through the exertions of the Macedonian horse. Yet their manoeuvring at this battle was not in keeping with the spirit of cavalry actions,
for their infantry kept pace with and engaged the enemy at the same
time and together with the horsemen.

Better tactics were displayed three years later at Gaugamela, when
the Macedonians had gained experience from the enemy. At this battle
the cavalry, 7000 strong, was formed in two separate bodies, and placed

on the flanks of the army. The cavalry of the right wing, led by Alexander in person, manoeuvred against the left wing of the Persians (composed of horse). Each tried to outflank the other, and the Persians had apparently succeeded in this, from their greater numbers, when Alexander suddenly deployed his deep columns to the right, and outflanked the enemy. In their hurry to prolong the left, to oppose this unexpected lengthening of the Grecian line, the Persians left an opening in their centre. Alexander at once dashed in, divided their forces, defeated and pursued them off the field. During the pursuit a message was brought to him from Parmenio (who commanded the left wing) asking for assistance. Alexander at once recalled his cavalry from the pursuit, and led them towards the rear of the Persian right. Finding, however, that Parmenio had already defeated the enemy with the assistance of the Thessalian horse, he turned about and followed up Darius. After crossing the Lykus, the men were allowed to rest till after twelve o'clock at night; then, resuming the pursuit, they arrived next day (the day after the battle) at Arbela, a distance of 600 stadiums!!* Here was the rapidity—the *dash*—which ought to characterise cavalry operations.

The Persian horsemen greatly outnumbered their foes in this battle, and they were clad in armour; yet they could not resist the close array and speedy advance of the Macedonians. Alexander had his horse well in hand, for in the midst of a victorious pursuit he suddenly recalled and led them to the support of the left wing, and their untiring pursuit of the enemy after all the fatigues of the battle proves that they were capable of the greatest exertions. With all this, after Alexander's death, the Grecian cavalry sank into its former insignificance.

With the Romans, cavalry achieved even less than with the Greeks; but under Hannibal the Carthaginian horse did wonders in Italy and on the soil of Rome.

They first met the Romans in a cavalry skirmish by the Ticinus. The Roman skirmishers (who were on foot) were speedily driven in, upon which they ran to the intervals, frightening the horses. Some of the Romans were thrown, others dismounted to fight on foot (a common practice amongst them). Whilst the cavalry were engaged, Hannibal sent

*Equal to about seventy-five English miles.

the Numidian light horse to turn the flanks of the enemy and attack him in rear. This gave him the victory. The Romans were defeated, and their commander, P. Cornelius Scipio, was wounded.

At the battle of the Trebia the Romans had 36,000 foot and 4000 horse; the Carthaginians 20,000 foot and 10,000 horse.

How different the proportions of each arm in the two armies!

The Roman infantry fought gloriously at this battle, as it ever did; their cavalry fled before that of the enemy.

The Roman Legions, then attacked on all sides, still succeeded in cutting their way through the enemy, though 10,000 only of these unconquered men reached Piacenza; the remainder were left on the field.

The Carthaginian horse won the battle of Cannae (B.C. 216). The Romans brought 80,000 foot and 6000 cavalry into the field: Hannibal had 40,000 foot and 10,000 horse.

The Roman right wing rested on the Aufidus: both armies had their cavalry on the flanks.

Hasdrubal first attacked the Roman horse with his cavalry, and drove them into the river. The battle now raged along the whole line. The Roman infantry, as usual, were everywhere victorious. The Numidian cavalry on Hannibal's right were engaged in a doubtful contest with the enemy's horse opposed to them.

Hasdrubal, who had done his work on the left, suddenly appeared on the right, defeated the cavalry, and, after sending the Numidians in pursuit, threw himself on the conquering Roman Legions, and, in spite of their heroic efforts, burst in amongst them, and defeated them with fearful slaughter. Æmilius Paulus and more than 40,000 Romans were slain, and most of the survivors made prisoners. Polybius gives the loss of the Romans at 70,000 men, but attributes their defeat to the fact of the Carthaginians being so superior in cavalry; and he adduces this battle in proof of his assertion "that it is better to have only half the number of infantry, if you are superior in cavalry, than to be on a perfect equality of all arms with the enemy."

The wide open plain, now called the Table of Apulia, on which this memorable battle took place, was admirably adapted to cavalry evolutions. It will be remembered how Æmilius endeavoured to persuade his rash and ignorant colleague Varro not to risk a battle there. After the combat Varro escaped to the near town of Venusium with only 70 horse. It is affirmed that in all this battle of Cannae, in which he made such use of that arm, Hannibal lost only some 200 horse. It was mainly through his cavalry, and his skill in handling it, that this truly great commander, generally cut off from all supplies from Carthage, was enabled to maintain himself in Italy for nearly sixteen long years.

The Roman Legions were so weak in cavalry that when successful against the Carthaginians they could reap no advantage from their victories; nor could they procure provisions and forage for their armies whilst the country round them was swarming with the enemy's horse. The most determined bravery on the part of their infantry could not always save them from defeat, and every affair in which they were worsted endangered their very existence.

Montecuculi says, "The principal act of an army is to fight a battle: this generally takes place in the plains, and then the cavalry is the most important arm engaged. For if the cavalry is defeated, the battle is generally lost; if, on the contrary, the cavalry is victorious, not only is the battle safe but the defeat of the foe is complete!"

Hence, also, Marshal Saxe's advice is to shut yourself up and temporize when you are weak in cavalry.

Fabius and L. Plancus, not daring to appear in the field against Hannibal's cavalry, kept to the hills and entrenched themselves.

In the same way, in the Thirty Years' War, Gustavus Adolphus would not venture on the vast plains of Poland, but remained in Prussia till reinforced in cavalry.

The want of cavalry stopped both Alexander and Caesar in their career of conquest.

The Romans afterwards swelled their cavalry at the expense of their infantry; but this was at the period of their rapid decline, when discipline, martial spirit, and patriotism, with every other virtue, were dying out among them. As the ancient and cherished Legion had become worthless, excellence was not to be expected from the mounted soldiers.

The estimation in which cavalry has been held has varied according to times and circumstances. There have been periods (and long ones) in which it has been prized above all infantry, and there have been others in which it has been considered as almost valueless when compared with infantry. Of course both these extreme opinions are wrong. Great battles have been won by each arm. But it is foreign to our purpose to balance the merits of horse and foot It is now admitted that cavalry must form a part of every army; and it is our business to make it as good in its way as our infantry is generally admitted to be.

History sufficiently proves the necessity of having a powerful body of horse with an army; and also shows that, if cavalry has not been at all times equally efficient or equally successful, it has almost invariably proved irresistible when well organized and properly led.

At Capua, A.D. 552, the Franks defeated the Roman infantry, but, being deficient in cavalry, they were outflanked and attacked by the Byzantine horse under the Eunuch Narses; and, according to the historian Agathias, out of 30,000 men composing the army of the Franks, five soldiers only escaped from the slaughter which ensued.

At the memorable battle of Poitiers, fought A.D. 732 by the Franks under Charles Martel against the Saracens under Abd-er-rahman, the cavalry of the Franks, led by Eudo Duke of Acquitaine, defeated the Moors and entered their camp, doing great execution. Paul the Deacon (Paulus Diaconus) says 375,000 Saracens were left dead in the field; other historians, amongst them Mezeray, estimate the strength of the Moor at 100,000, and assert that most of these, including their king, were killed or trampled under foot by the victors. But, in spite of these assertions, the Saracens left their camp in open day and retired after the battle across the Pyrenees without being molested by the enemy. The cavalry was not sent in pursuit, from which we must infer that they could not move with sufficient celerity in those days to follow their more nimble opponents with any chance of success.

In 933, the battle of Merseburg, the European cavalry, under Henry I, had greatly improved in organization and tactics. By their compact order and discipline they gained a decisive victory over the Hungarian irregular horsemen, who at that time were greatly dreaded in Europe.

The success of the cavalry was here due to the manner in which Henry had re-organized and trained them; and it is worthy of remark that light horsemen are first mentioned in this battle. Armed with crossbows, they competed successfully with the Hungarian horsemen, and distracted the enemy's attention during the battle by constant skirmishing.

The Magyars, 300,000 strong, left their entrenched ramp near Schkölzig and advanced into the plains N.E. of Lützen to meet King Henry. The battle was long undecided, till Henry, breaking forth at the head of his cavalry, which had been concealed near Schkölzig, attacked them in flank and routed them.

The enemy was at once pursued, nor was any respite given to the fugitives till they had crossed the Bohemian frontier.

History affords but two examples of cavalry being thus successfully re-organized and well commanded in the field, namely, under Henry I in the tenth century, and under Frederick the Great in the eighteenth century. Both these princes on their accession, found the cavalry badly organized, and in their first campaigns worse than useless; but they reformed and achieved with them their most brilliant victories.

King Otto I, in *the field*, managed well the cavalry which had been re-organized under his father Henry. At the battle of Augsburg, in August, 955, he formed them into separate corps to support each other; and whilst he poured his heavy horsemen in great numbers on one point to strike a decisive blow, as soon as he had gained an advantage he sent his light horse in pursuit and, gathering up his mail-clad horsemen, struck again and again, till victory crowned his efforts. The battle was lost and won more than once during the day; for the Hungarians, though their ranks were open and their files scattered, yet, like all good light cavalry, worked well together in their attacks, as if each individual horseman was instinctively animated with the same intention. Like shadows they eluded the grasp of the heavy Germans,—never left them,—and hung on their heels when they turned back again. The better generalship of King Otto alone gave him the victory.

We now go on to the thirteenth century, when the Mongolian hordes invaded Poland; and we find, at the battle of Liegnitz (9th April, 1241), that the Polish cavalry, all heavily equipped and partly clad in armour, were defeated by the more active horsemen of the Mongols.

The armies met in the plains near Wahlstatt. The Polish army, under Prince Henry the Good, and Mizeslav Prince of Upper Silesia, were marshalled in five corps, of which some in reserve: their opponents formed much in the same way. The Polish right wing attacked and defeated the Mongols opposed to them, but these soon rallied and drove them back on their reserves, which, in their turn, advanced, overthrew the Mongols, and pursued them. Suddenly, however, according to Polish historians, "a spell, the effect of enchantment, began to work," and their countrymen fled. This spell originated in the quick rallying of the Mongols, and their sudden and unexpected return to the charge when their enemies were exhausted in trying to catch them. Such a surprise was very likely to act like magic upon the unwieldy horsemen of the West.

Prince Henry himself now advanced with a corps which had not been engaged: Peta, the leader of the Mongols, came to meet him, but, as before, evaded the shock of the first onset, and returned again and again to the charge; when his adversaries were fairly tired out, he then, with a fresh reserve (the last one in hand), poured down and swept the worn-out Christians from the field.

Here, as in the other instance, the Poles followed up their first success, pursuing the Mongols when they purposely retired before them; again they were spell-bound; and this time it is said that, wrapped in clouds of dust, they were deprived of sight and strength.

We understand this "spell" also, and so did the Mongols; for, as soon as they saw it working, as one man they turned on their Christian pursuers, and slaughtered them without mercy.*

The contact with Eastern nations has, from the earliest times, influenced the progress of cavalry in Europe; and the very tactics displayed by these Mongolian hordes in the thirteenth century are the rule and the foundation of our cavalry tactics of the present day.

The improvement of that noble animal, the horse, is comparatively of recent date in England. At least as late as the time of Charles I we looked to the Continent—chiefly to Spain and Naples—for our best horses for the manège and war. Our Edward I took cavalry with him into Scotland; and we know what became of the English horse at Bannockburn. The battles in France of our Edwards and Henries were infantry battles, and were gained almost entirely by the English bow. The

---

*Kriegswissenschaftlichen Analecten, by Capt Ganzauge.

French had, at times, a numerous cavalry, but it was deplorably deficient in discipline; and when they had a few good squadrons they evidently knew not how to use them. Good horses were scarce, and they had neither the patience nor the money to train the mounted soldier as he must be trained ere he can be serviceable in war. The knights, and some of the regular men-at-arms, rode well, from being constantly in the saddle; but the rest (as we have seen it said of certain modern horse-soldiers raised in a hurry), "not being accustomed to horses, were always falling off." It takes threefold more time to teach a man to ride and to have a perfect mastery of his horse than it takes to teach a foot-soldier his complete drill, and when the horse-soldier is taught so far, he has still a vast deal to learn before his education can be considered as completed. This difference alone will go far to show how it has happened that the cavalry soldier has been inferior in his line to the infantry soldier in his.

In modern Europe cavalry first rose into importance when the nations of Germany overran the Continent, and felt the necessity of having numerous bodies of horse with their invading armies; a necessity which had not been made apparent whilst they remained in their own country.

By degrees European cavalry were clad in armour, formed in single ranks, and each horseman was expected to single out an antagonist for the fight.

The select French gensdarmes, who were cased in heavy armour, greatly distinguished themselves in this manner of fighting.

To these heavy horsemen succeeded the Reiters, German mercenaries, who, mounted on faster horses, equipped more lightly, and armed with swords and pistols, constantly beat the gensdarmes in the civil wars of France and Flanders.

The introduction of gunpowder had brought about a change; and the cavaliers by degrees had laid aside their armour, and taken to fire-arms.

Under Henry II of France cavalry was formed again in oblong squares and in ten ranks. Henry IV reduced their front and, by degrees, the number of ranks to six.

In the sixteenth century detachments of Infantry were mixed with the cavalry; and both made to keep pace, and meet the enemy together in battle.

At the battle of Pavia, in 1525, the Marquis of Pescara placed (for the first time) bodies of heavy-armed Spanish musketeers in the intervals of the Imperial cavalry, and *thus* (it is said) defeated the French.

The practice of mixing horse and foot gained ground; and as the infantry soldier rose in importance, so did the cavalry soldier sink lower and lower in the scale.

With the introduction of gunpowder the distance at which contending armies engaged in battle was greatly increased: cavalry could effect nothing with fire-arms, and were kept out of the reach of the enemy's shot; they were of little use, because their movements were so slow that the opportunity for action was generally lost before they could go over the ground which separated them from the enemy.

Gustavus Adolphus reduced the unwieldy size of the squadrons at the beginning of the Thirty Years' War (1630); he formed them in four ranks, of which the first three charged whilst the fourth remained in support; he stripped the men of their armour, took from them their lances, and made them lighter, more active, and more useful. He formed his cavalry in two lines, and generally placed them on the flanks of the army.

In France, under Louis XIII, in 1635, squadrons were reduced In front and depth, and in 1766 they were formed in two ranks; they were still unwieldy and difficult to manoeuvre, though they had improved by being organised in regiments, and having the heavy equipments made lighter.

The practice, however, of sending cavalry into action supported by bodies of musketeers placed in the intervals, took from them their impulsive power, and deprived them of the advantage to be derived from the speed of their horses.

It was reserved for Charles XII to alter this style of fighting on horseback, and to improve upon it.

His daring and chivalrous character was suited to the spirit of cavalry tactics; he led his horsemen sword in hand against cavalry, against infantry, against fortified positions, over any country; he acknowledged no difficulties and overthrew all opposition. Untiring in pursuit, he actually followed up the Saxons under Marshal Schulenburg in their retreat into Silesia for nine consecutive days without unsaddling, overtook them at Sanitz, near Punitz, and, with two regiments of cavalry only, charged them, though ten thousand strong—rode over their infantry, who lay down to avoid the impetuous rush of the Swedes—defeated and drove the Saxon cavalry off the field—then returned to attack the infantry and guns. Night alone put an end to the combat, and the enemy profited by the darkness to escape across the frontier. All the guns fell into the hands of the Swedes. They appear to have made a most formidable use of their long straight swords when in pursuit of the cavalry, for the dead Saxons had all been run through the body.

The period of our Great Civil War witnessed the introduction of many changes besides those which were merely of a political nature. It was then that our horse first began really to distinguish themselves, and to stand forward as the gainers or deciders of victory.

The English cavalry under Cromwell and his fiery adversary Prince Rupert claim especial notice; for from the numerous cavalry engagements of that period many good and useful lessons may yet be gleaned by the cavalry soldier.

Cromwell, forty-four years of age when he first drew a sword, showed himself a great soldier at the very outset. He himself raised, organised, and disciplined his troops of horse, and set his men an example which.they were not slow in following. His mental and bodily energy, his vigorous conceptions, quick decision, and the dread vehemence with which he urged his war-steed into the thick of battle, made of him a cavalry leader second to none in history. Indefatigable and active, a good horseman, and perfect master of the broadsword, he had unbounded ascendancy over the minds of his followers, and led them through, or over, all obstacles that human prowess could surmount.

The impetuosity and rashness of Prince Rupert were no match for the cool courage and presence of mind of Cromwell. The latter often turned defeat into victory; the former lost many a fair field by letting his cavalry out of hand after a first success; and, during his absence, his able opponent secured the prize.

At Grantham the Royalists had one-and-twenty troops of horse, and

*The troop consisted of sixty men, one captain, a lieutenant, a cornet, and a quartermaster.

three or four of dragoons.*

Cromwell drew out about twelve troops; to meet them; they formed at musket-shot distance from each other, and the dragoons fired for half an hour, or more. Cromwell then led his troopers on to the charge, sword in hand; the Royalists received him standing, and were at once overthrown; he followed them up for some miles, each trooper killing two or three men in the pursuit.

At Gainsborough, after a skirmish with the enemy's advance, Cromwell gained the crest of a hill, and saw suddenly a great body of the Royalist horse facing him, and close to him, with a good reserve of a whole regiment of cavalry behind it. Though taken by surprise, Cromwell led on his horsemen to meet the foe, who were pressing forward to take him at a disadvantage. A good fight with sword and pistol ensued, till the Roundheads, pressing in upon their adversaries, routed the whole body, and immediately pursued, doing execution upon them for five or six miles.

Cromwell, however, who commanded the right wing, kept back Major Whalley and three troops of horse from the chase; these he at once formed up, and, observing the enemy's reserve, under General Cavendish, charge the Lincolners and rout them, he suddenly galloped in on his rear, drove horses and men off the field at the sword's point, and killed Cavendish.

At the famous and disastrous battle of Marston Moor, fought on the 2nd of July, 1644, Cromwell signally distinguished himself, and gave Prince Rupert a taste of the steel of his Ironsides which the latter did not at all relish.

A junction had been formed between the Scotch army and the English Parliamentary forces, and they invested York. Prince Rupert and the Marquis of Newcastle joined their forces to raise the siege, for the possession of this ancient city was of great importance to them in a military point of view.

The opposing forces numbered about 50,000; they were drawn up with a ditch between them, and did not get into position till five in the evening; the king's troops facing the west, their opponents the east. A long and bloody contest then ensued. At first neither party would give up the advantage which the dam and ditch afforded to those who remained on the defensive, till Lord Manchester moved forward with the left wing of the Parliamentary army to the attack, seconded by Cromwell, who commanded the cavalry of that wing.

The attack was successful at every point, though a desperate fight took place between Cromwell's and Rupert's horsemen. Cromwell had kept part of his cavalry in reserve; these suddenly fell upon the Royalists whilst engaged in the *mêlée* with him, and completely defeated them.

The right wing of the Royalist was now closely pursued by horse and foot, and driven far back behind the left wing.

The exact counterpart to this had taken place on the opposite wings of the contending armies.

The left wing of the Royalists had advanced, attacked, and driven hack the right wing of the Parliament, defeating their horsemen, who, in galloping to the rear, spread confusion and dismay amongst the reserves of Scotch infantry.

Lord Manchester only heard of what had happened on his right after Sir Thomas Fairfax's troops had fled some miles on the road to Tadcaster; and Cromwell, at once collecting his cavalry from the pursuit, turned and followed the victorious Royalists towards that place. These formed to receive him, were defeated, and fled. Thus Cromwell, by his energy and courage, won the day, after some of the chief generals had left the field, and given the battle up as lost. The battle of Marston Moor resembles in many particulars that of Zorndorf; both were brought about in the same way, both were gained by the same manoeuvre, which Hasdrubal employed at Cannae, and Seidlitz at Zorndorf.

These three great cavalry generals were victorious first on the left wings of their armies; from there they passed to the right wing to reestablish the fight: and all three then succeeded in the same way in breaking through the enemy's infantry, which, again, in all three cases fought manfully. Thus these three great battles are reckoned as cavalry victories, for to the horsemen was due the success on all three occasions in question.

Curious that at such distant periods, under such different circumstances, the same results should have been achieved, and in almost exactly the same manner! Does this not point forcibly to the necessity of consulting the past, that we may prepare for the future?

On the 14th of June, 1645, was fought the battle of Naseby. The right wing of the Royal army was under the gallant Prince Rupert, the left under Sir Marmaduke Langdale, the main body under Lord Astley, the reserve under the King in person.

Of the Parliament, the right wing was commanded by Cromwell, the left by Ireton, the main body by Fairfax and Skippon. Rupert charged hotly the left wing of the Roundheads, fairly broke them, drove them through the streets of Naseby, and continued the pursuit. Cromwell, at the same time, charged and dispersed the Royal horse under Sir Marmaduke Langdale. The Royal infantry. In the mean time, engaged and were driving the Parliamentary foot before them. The fate of the day depended upon which side should first see their cavalry return. Cromwell at last appeared, at the head of his dreaded Ironsides, charged fiercely on the flank of the Royalist infantry, and threw them into irremediable confusion. The fate of the day was sealed: Rupert returned too late to do any good; and the King fled the field, leaving his artillery and 5000 prisoners in the hands of the victors.

Thus had Cromwell again turned defeat into victory by throwing his Ironsides into the scale at the proper time; whilst Rupert, by his headlong and thoughtless bravery, lost to his sovereign for ever the chance of recovering his crown and kingdom. Had he returned after his first success and attacked the Roundhead infantry, the issue of the whole struggle would probably have turned in favour of the King, for it required the best efforts of Fairfax and Skippon to prevent their infantry from running away.

A few of the officers who served in this war had fought under Gustavus Adolphus, and had charged with the Swedish cavalry in the plains of Germany, and on the field of Lützen, where the hero fell.

For ages the finest cavalry seen in Europe was indisputably that of the Turks. In great part, both men and horses were brought over from the Asiatic provinces of the empire, and the rest of the men and horses were principally of Asiatic descent. The horses, though not large (seldom much exceeding 14 hands), were nimble, spirited, and yet docile, and so trained and bitted as to be perfectly under control: the hollow saddle was rather heavy, but all the rest of the appointments were light: the soldier rode in the broad short stirrup to which he and his ancestors had always been accustomed, and on which they had a firm and (to them) natural seat out of which it was most difficult to throw him; his scimitar was light and sharp, and, in addition to it, he generally carried in his girdle that shorter slightly-curved weapon called the yataghan, with an edge like that of a razor. Some of the Spahis used long lances or spears, but these were always thrown aside, as useless, in the *mêlée* of battle. Their tactics were few and simple. If they could not get in the small end of one wedge, they tried another and another wedge; if they penetrated the hostile line, they dealt death around them, their sharp weapons usually inflicting mortal wounds or lopping off limbs. If the enemy gave way, they spread out like a fan, and, while some pressed on the front, others turned the flanks and got into the rear. Occasionally, to gain time, the Turks mounted some of their infantry, *en croupe*, behind their Spahis. Thus, early in the battle of Ryminik, when they had to contend with Marshal Suwarrow and some Austrians, a body of 6000 Janissaries jumped up behind an equal number of Turkish horsemen, and were carried at full speed to occupy a commanding eminence, of which the Austrians were also desirous of taking possession.*

We have seen, quite in our own day, this effective and really brilliant cavalry reduced, by the spirit of imitation and ill-understood reform, to a condition beneath contempt. The late Sultan Mahmoud must needs have his cavalry disciplined *alla Franca*, or in Christian fashion: and he imported a number of French, Italian, and German non-commissioned officers to teach his men to ride with long stirrups, and to form, dress, and look like Europeans. To the disgust and even dismay of his Moslems, he buttoned them up in close jackets and put them into tight pantaloons. With a most perverse determination the system has been continued and extended these last twelve years, under his son and successor, the present Sultan Abdul Medjid, and it may now safely be said that the Turkish cavalry is no longer the best in the world. The men, always accustomed to sit cross-legged, and to keep their knees near the

*Marshal Marmont, *Travels in Turkey*, &c.

abdomen, cannot be taught to ride with the long stirrup, *à la Française*. They are always rolling off, and are frequently ruptured. They are armed with the lance, and have seldom any other weapon except an ill-made, blunt, awkward sabre. Their horses are now wretched *rosses*. The good breeds have died out; and the Imperial, centralizing tyranny— masked under the names of reform and civilization—which has been raging with more or less intensity these last fifty years, has not left on the surface of the empire a man of hereditary rank and wealth, or any private country gentleman, with the means of restoring the lost breeds, or of supplying such good light cavalry horses as existed in abundance at the commencement of the present century. The Karasman Oglus, the Paswan Oglus, and all those great Asiatic feuda-tories, together with the hereditary Spahi chiefs of Roumelia, who kept up the principal studs, are all gone! Mounted as they are, armed as they are, and riding as they do, instead of dealing with European horsemen after the summary fashion of the good old Turks, any English hussar ought to be able to dispose, in a minute, of half a dozen of Abdul Medjid's troopers, trained *alla Franca*, though he (the hussar) were armed only with a stout walking-stick.

The cavalry of the Russians and Austrians improved much during their wars with the Turks; and of the knowledge thus practically obtained the Austrians made good use in their first campaigns against the Prussians in the Seven Years' War.

These Turkish horsemen, without discipline, rushed in like a whirl-wind, in swarms or irregular columns, and swept over all that came in their way, leaving death alone in their track, so effectually did they ply their scimitars.*

Neither discipline nor the fire of artillery and infantry would save the Christians from these fanatic horsemen; their only safety lay in the chevaux-de-frise with which every column was provided, and each bat-talion had two light carts to carry them.

When in the neighbourhood of the enemy the men took the chevaux-de-frise on their shoulders, formed a skeleton column, and, when an attack was threatened, they wheeled into line, fixed the joints in the ground, and fastened them together.

*Berenhorst.

To these arrangements the Russians owed their first success against the Turks, as far back as 1711.

When General Munich marched out against the Turks in 1736, he did not consider the chevaux-de-frise a sufficient protection, and again armed part of his infantry with long pikes. His troops marched in large oblong squares; these were at a moment's notice surrounded by the iron spikes of the chevaux-de-frise, and flanked by artillery. At this impassable barrier they received their turbaned assailants, and poured upon them a destructive fire in perfect safety.

No European cavalry, with all its tactics, large squadrons, cuirasses, and lances, ever inspired such dread, or brought infantry to the necessity of seeking safety behind impassable obstacles. The Moslems alone inspired sufficient dread to call forth on the part of the infantry a humiliating confession of their weakness in the precautionary measures they adopted; for, unless surrounded by these formidable engines of war, the Turks seldom failed to burst in amongst them, and then handled the sword quick, masterly, and without cessation, until checked by the reaction brought on from the excess of their own fury.*

With European cavalry they dealt in the same summary way whenever they got amongst them: but, to prevent this, the cavalry were formed In masses, with guns and infantry on their flanks!!

Now, if the individual prowess and skill in single combat, the horsemanship and sharp swords of the Turks, made them so formidable as history here relates, how irresistible would cavalry be which to these qualities should add that discipline and method in which they were wanting, and which was the cause of the disastrous termination of all their wars after the close of the seventeenth century!

The Mamelukes of Egypt kept up their high qualities as bold horsemen until they were annihilated at the commencement of the nineteenth century: but they can scarcely be said to have belonged to the Turks. If these brave Mamelukes, drawn from different races and from different countries, but chiefly from the ancient Thessaly and Macedonia, and from the backgrounds of European Turkey which we now call Servia, Bosnia, Albania, etc., had been backed by only a tolerable infantry, the sanguinary affair at the Pyramids would have been a

*Berenhorst.

defeat and not a victory to the French. Single-handed, the French troopers had no chance with those daring horsemen and expert swordsmen.

While the Russians and Austrians were impelled by the Turks into an improvement of their cavalry, pains were taken by the Prussians to add to the efficiency of that arm. Wherever there was war, or a probability of it, it was seen and felt that cavalry must bear an important part, and that there was much to change or to modify in it. Nobody thought that, while infantry and artillery were improved, cavalry could be left *in statu quo*.

Frederick William the Stadtholder and Leopold of Dessau together reorganized the Prussian army, and laid the foundation of that discipline which, under Frederick the Great, became so celebrated, and was copied by almost all European nations.

Frederick William would have tall men for his army: they were kidnapped by his recruiting parties wherever they were met with.

His cavalry were well drilled to fire in line, both on foot and on horseback: nothing was done to make them formidable in close combat; they charged at a walk or a trot.

When Frederick the Great ascended the throne, he found his cavalry drilled in this way. The horses and men were colossal: they dared not *walk* on a bad pavement, or move beyond *that pace* on uneven ground.

At the first battle against the Austrians, the Imperial cavalry, which had gained experience in the Turkish wars, charged the Prussians sword in hand, Moslem fashion, at speed, and drove them from the field. The Great Frederick, who did not like the look of matters (at this battle of Mollwitz), took the advice of his field-marshal, followed the fugitives, and only rejoined the army next morning, on hearing that his infantry had stood firm and won the day in spite of the flight of the cavalry.

When the campaign closed with the conquest of Silesia, Frederick at once proceeded with the organization of this arm of the service. He began by doing away with all firing in line, and gave all his attention to making them good riders. Seidlitz formed his hussars in two ranks, and towards the end of the campaign the remainder of the cavalry followed his example. They were all brought to do what Marshal Saxe laid down as necessary, namely, to charge at their best speed for 2000 yards without breaking their array. Many of the old Prussian generals opposed these innovations to the utmost; but the King carried them through, for he was convinced of the advantage of impetuosity in the attack: and his mounted troops, which had been defeated constantly in the first campaign of the Seven Years' War, when thus reorganized and led by Ziethen and Seidlitz, astonished the world by their deeds of arms; not only overthrowing cavalry in their headlong career, but sweeping whole armies of infantry off the field. Witness the battles of Strigau, Kesseldorf, Rossbach, Leuthen, Zorndorf. This last was the most glorious of all to the Prussian horsemen, who, in thirty-six squadrons, under Seidlitz, not only turned the fortune of the day, saved the infantry and artillery of their own army, but checked the advance, overthrew the victorious Russian cavalry, driving it from the field; then returned to fall upon the Russian infantry, which, prepared to receive the Prussians, fought with the most determined bravery; and when their masses were broken into by the furious horsemen, those who escaped the sword threw themselves again into masses, and had to be charged again and again. In no modern battle did so many men fall by the sword as at Zorndorf, though the Prussians had been twelve hours on horseback before advancing to the charge.

At no time have more glorious deeds been done by cavalry than were achieved by the Prussian horsemen of those days. Their arm was the sword: their trust lay in the individual prowess and good riding of their horsemen; their tactics consisted in speed and determination: and to this system is attributed not only their wonderful success, but also their generally trifling loss in killed and wounded.

Seidlitz practised his hussars at going across country, using their swords and fire-arms at speed; and various were the feats to which he drilled his men, in order to make them expert in the management of their horses and arms.

An anecdote is related of him. When the King inspected his regiment and found fault with the number of deaths occasioned that season by accidents at drill, Seidlitz answered very drily—"If you make such a fuss about a few broken necks, your Majesty will never have the bold horsemen you require for the field." It was one of the amusements of this daring cavalier to ride in at speed between the arms of a windmill while working. This feat Seidlitz often performed after he had attained the rank of a general officer.

The ancient Greeks followed much the same system as Seidlitz, being convinced that neither man nor horse would be up to the work unless frequently put to it beforehand. "If you wish to have a good war-steed," says Xenophon, "you must try him in all those things which may be required in war. These are, to leap across ditches, scramble over walls, spring up ascents and dash down descents; and to be experienced in charging on slopes, uneven ground, and transverse roads or paths. Many horses fail, not for want of ability, but for want of experience in these things. Let them be instructed, trained, and accustomed, and they will excel in them all, if they are healthy horses and not vicious."*

Frederick the Great divided his cavalry in the field into corps of twenty, thirty, and forty squadrons, and made them stand out boldly and alone to play their own part according to circumstances. He wielded sword and sceptre. He directed these cavalry movements with consummate skill and energy: he let no opportunity pass without making his enemies feel the weight of his sword; and the Prussians, thus encouraged by their King, and full of confidence in their leaders, plied their spurs and rode to victory.

It was a favourite saying of Frederick that three horsemen in the enemy's rear do more than fifty in front; and his generals always tried to attack front, flank; and rear at the same tune. In the two first attacks, or in front and flank, they generally succeeded. How they did so has remained a mystery to this day. It however appears that they generally seized the moment when the combined use of artillery and infantry, or the use of either singly, had made an impression, and then dashed in; or that they rapidly gained the enemy's flank and charged home. Out of *twenty-two* great battles fought by Frederick or his generals, the cavalry, thus employed, decided in the fate of *fifteen*.

To his cavalry in action Frederick gave no orders beyond general directions as to which part of the field it was to act in. The moment for attack was always left to the generals commanding the cavalry, who, after securing their flanks and providing a reserve, spurred and started;

*Xenophon, *On Horsemanship* [3.7].

and, being once started, they pushed on whilst there was an enemy in the field.

Berenhorst (in his *Betrachtungen über Kriegskunst*) gives the following interesting account of the battle of Rossbach:—

"The generals were dining with the King at Rossbach, when a cry arose of 'The French are coming!' They jumped on their horses, and, as if by inspiration, gave the order to fall in, form column, and take ground to the left; they must have been beaten had they awaited on their own ground the well-planned attack of the enemy; but, without any intention of misleading the French, they left the tents standing for they had no time to strike them; and this accidental circumstance deceived the enemy better than the most cunningly devised scheme.

"The right wing of the Prussians stood fast; the other, marching in column by its left, and screened by a rising ground from the view of the French, gained their flank, whilst the enemy, advancing to surround the Prussians, suddenly hesitated. Revel (one of the Broglio family), who led the French attack, fell mortally wounded at the first discharge, and left his columns with their flanks exposed, and not knowing what they were to do next.

"The Prussian army, numerically weak, were full of ardour, the forerunner of victory: they despised the French. Those who were moving towards their flank and rear looked upon the whole thing as a good joke; they were delighted at the idea of catching the enemy at a disadvantage, and in this they succeeded.

"The *genius* of the Prussian cavalry sprang forth here from the fields of Reichardtswerben, and led them to victory.

"When the cavalry in order of battle, like a pent-up flood, is held ready, and, at the first signal, poured down in torrents, floods the field, sweeping all before it, then has cavalry reached the ideal of perfection; and to this ideal Seidlitz attained with the Prussian cavalry on that day. Soubise and Hildburghausen were swept from the earth."

In reviewing the deeds of the Prussian cavalry of those days, it must be borne in mind that they dealt with infantry, which sought the open plain, advanced in long lines—avoiding obstacles of all descriptions, because such obstacles disturbed their array. Their fire was quick, but not true in its aim, and their squares seldom held out long against the horsemen.

In those days an individual could often take in at one glance the whole state of affairs at any time during a battle, and thus employ the cavalry at the proper moment. But with the improvements in fire-arms the extent of ground occupied by armies in position has gone on increasing, and to such an extent that it is no longer possible to overlook the field, and, therefore, more difficult for a cavalry leader to achieve the same results as the Prussians did under Seidlitz and Ziethen.

Cavalry must now act more in unison with other arms; for great results are now achieved only by their skilful combinations.

Since the Seven Years' War cavalry has fallen in general estimation, and has lost that proud preeminence at which it stood when it decided the fate of battles. Many gallant deeds have been done in later days by English and German horsemen, as at Avesne-le-Sec, Villers-en-Couche, Cateau Cambresis, Emsdorf, Usagre, Salamanca, Garci-Hernandez, and Waterloo; and by the French in many a well-contested field; but they do not come up to the exploits of the horsemen of Frederick the Great, who, held in hand in large numbers, till the opportunity offered or necessity required them to be let loose, then burst over the battle-ground, and swept down all in their impetuous course; the word on their hearts, as well as lips, being "Charge home!"

At the commencement of the great war of the French revolution (1792), the cavalry of our neighbours was very far from being either numerous or good. In fact, as a nation, the French are not, and have never been, truly equestrian. Generally, they are bad riders, and without good riding there can be no thoroughly good horse-soldier. We do not think that this deficiency is accounted for by the fact that in nearly all parts of France the ground is tilled, not by horses, but by oxen. We attach more importance to a second reason assigned by General Foy: the Frenchman's impatience, or what the General calls *vivacité inquiète*, may prevent him from identifying himself with his horse, or from learning to ride as he ought. He has, besides, an hereditary superstition for the long stirrup, and for balance-riding, which never yet carried a man across a rough country without disaster. In their first campaigns the French had little chance against the German heavy horse, the Hungarian hussars, or even the Walloon dragoons. They seldom presented much cavalry in the open field, and when they did it was usually to their disadvantage. Moreover, the French horse were poor, under-sized, and under-bred. They got better remounts when they conquered other countries with their infantry, artillery, political propagandism, and daring strategy. But the war-horse is nothing without the rider, and cavalry soldiers are not to he improvised quite so fast as foot-soldiers.*

Before the reign of Bonaparte some regiments of heavy cavalry served as a corps of reserve to each army, the rest of the horse being scattered among the divisions of infantry or joined with the artillery. Napoleon tried to give his cavalry the same part to act in battle as Frederick the Great had given to his; but he organized them differently: and widely different were the results. Napoleon's horsemen were not at home in their saddles; they were heavily equipped, and could not move with speed: he therefore formed them into very large masses, which obtained the curious name of *Corps d'Armée de Cavalerie*. In these large corps he attached guns to each regiment, and used deep formations for attack; thus his cavalry played a secondary part to the artillery; its movements were cramped, its approach necessarily slow, and, as it was always heralded by its own cannon, the enemy was seldom taken by surprise (except at Marengo), but had time to prepare a reception which cost the French masses of horse very dear. Still his horsemen, mostly clad in defensive armour, were poured on slowly but in irresistible numbers, and thus, regardless of the loss of life, Napoleon by their means won many a field. Even allowing for all the brilliancy of Murat, it may be doubted whether he had one cavalry leader whom Frederick the Great would have called good.

Napoleon's cavalry generals often failed in bringing their troops into action at the right time, and often threw them too early into the scale, and so, when a reserve of cavalry might have decided the fate of battle, none was forthcoming.

---

*"La cavalerie n'est pas si facile à improviser que l'infanterie." General Foy, *Histoire de la Guerre de la Péninsule*.

They often neglected to protect their flanks or to have a reserve at hand in case of disaster. In 1813, on the 16th of October, near Leipzig, the cavalry corps of Latour-Maubourg and Kellerman, about 5000 horses, led by Murat in person, attacked the centre of the allied army advancing by Wachau towards Gossa, overthrew the division of Russian Light Cavalry of the Guard, captured thirty pieces of cannon, and broke through the line; but, 400 Cossacks of the Guard, gallantly led, fell upon their flank, and not only retook all the guns, but drove them back, in confusion, turning the whole affair to the advantage of the allies. These Cossacks had to gain the flank of the enemy by a path which admitted only of single files.

At the battle of La Rothière the same mistake, and in a greater degree, was again committed by the French.

The cavalry divisions, Colbert, Guyot, and Piré, having charged and overthrown the Russian division of hussars under General Lanskoy, were preparing to fall upon the infantry, when General Washiltschikoff brought up the Pantschulitscheff division of dragoons at a gallop, attacked the French in front and flank, drove them from the field, and pursued them to Alt Brienne, occasioning the loss of twenty-eight guns to Napoleon's Garde: yet they had plenty of cavalry in the field with which to have protected their flanks, but it only made its appearance after the affair was decided in favour of the allies.

Instances of this sort might be adduced of the English cavalry.

Charges, gallant and daring in their character, were turned into disgraceful defeats or dreadful losses by the culpable negligence of their officers in not having reserves in hand to protect the flanks during an attack, or to oppose an enemy coming on with fresh troops.

In the Peninsula, in 1812, two regiments of English horse under General Slade attacked and defeated two regiments of French dragoons near Llera, pursued them madly for about eight miles, when the French General, Lallemande, fell upon them with his reserves, and routed them completely.

The Union Brigade under General Ponsonby, at Waterloo, suffered severely from the same cause: after riding down everything in their way, entering the enemy's position, and sabreing the artillerymen at their guns, they were suddenly attacked by the French cavalry reserves, and driven back with great loss.

The 3rd Dragoons charged the Sikhs at Moodkee, and drove along the rear of the whole of their position: not only were they not supported, but our own artillery played upon them at one time, and occasioned them some loss. This gallant regiment returned to camp in the evening, having lost nearly two-thirds of their number in killed and wounded, and effected very little except inspiring a wholesome dread of English dragoons.

2

THE AUSTRIAN, RUSSIAN, AND PRUSSIAN
CAVALRY 1793–1815

L ET US HERE RAPIDLY EXAMINE the cavalry of Austria, Russia, and
Prussia, as displayed in action, from the year 1792 to the end of the
war of the Revolution.

In their first campaigns the cavalry of the allies surpassed that of the
French in every particular. They may not, in every case, have made the
best use of this advantage, but it is most indisputable that this superiority existed, and that the allies had it. Their cavalry corps were composed
of men essentially horse-soldiers by nature and habit, brave, numerous,
well mounted, and well organized: their individual superiority was
made apparent in every action from 1793 to the close of that century.
That the advantages they gained over the French in these numerous
engagements was never attended with any decisive result on the campaign, can be attributed only to bad generalship.

The French cavalry, at last, defeated the cavalry of the Austrians at
the battle of Hochstedt, in the year 1800; gained by degrees a complete
ascendency over all other continental cavalry; and, in spite of its many
inherent defects, contributed, in no small degree, to Napoleon's successes in the field, until their victorious career was buried with them in
the snows of Russia in 1812.

The cavalry of the Austrians and English, in 1793 and 1794, achieved
at various times the most brilliant successes in the Netherlands. Neither
French horses nor French men could stand against ours. It they met, the
weaker were literally rode down or rolled over; but, unluckily, our
horsemen knew little more of their *métier* than how to make a charge,
nor did they always know how to do that in the best way.

The campaign of 1795 (on the Rhine) was one of manoeuvre. The French were defeated at Handschusheim with the loss of two thousand men and ten guns, chiefly through the gallant conduct of the Austrian dragoon regiment Kaiser, and the hussars of Hohenzollern and Szekler. These hussars of the period have often been described by old soldiers of various nations (not excluding France) as the very perfection of light cavalry. The most interesting incident in the campaign was the storming of the lines round Mayence, on account of the glorious part taken therein by the Austrian cavalry. Three columns of attack were formed, and to each a few squadrons were attached; a reserve of twenty-two squadrons was kept in readiness, and the very moment the infantry stormed the lines these horsemen rode in, were let loose on the enemy, and achieved a complete victory, with comparatively little loss to themselves.

The French lost all their guns, amounting to one hundred and thirty-eight, and upwards of three thousand men; here one might well say with Suwarrow, "*Vivent le sabre et la baïonnette!*"

In the succeeding campaign of 1796 the Imperial cavalry did good service, and were well led during the battle of Würzburg, but were made no use of after they had gained the victory.

The whole of the French cavalry had been united at this battle under the command of General Bonnaud; the cavalry of the Austrians met them close to Emsfeld, and drove in their skirmishers; Bonnaud, seeing the Imperial cavalry gradually increasing in numbers, thought it best to charge without loss of time. The French fell on with resolution, and drove back the left wing of the Austrians, who retired on their reserves; in the mean time fourteen squadrons of hussars burst forth from behind a village, and galloped in on the rear of the French, who were simultaneously attacked in front by the German cuirassiers: the remainder of the French cavalry were then thrown in to the rescue, but the Austrians held twelve squadrons still in reserve, and these decided the fate of the day; the French were driven off the field, pursued behind their infantry, and two battalions of the division Grenier were afterwards destroyed by the victorious Austrians.

After the storming of Mayence the Imperialists rested on their laurels, and concluded an armistice; after Würzburg, instead of following up Jourdan and destroying his army, they halted, satisfied with the results of the day. They *stopped* at the point which Frederick the Great would have considered only as a glorious *starting* point.

Never was so fair an opportunity thrown away; under the fiery spirit of a Seidlitz the cavalry would have swept like a flood over the retreating and disorganised army, and made of Würzburg a second Rossbach for the French.

In 1797 nothing worthy of mention was effected by cavalry on the Rhine; but in Italy the Austrian cavalry distinguished itself at Mantua. They were encamped near the fortress. Napoleon, wishing to drive them into it, determined to attack. Massena's division succeeded in surprising them on the morning of the 14th of September, and had already entered the camp, when the Austrian cavalry, which had gone out in watering order, returned at a gallop, and (without saddles on the horses) fell upon the enemy and defeated him.

The allies (in 1805), who had 172 squadrons of cavalry at Austerlitz, knew nothing of the position of the French army nor of Napoleon's dispositions, which might have been easily unmasked by a "reconnaissance en force."

During the battle they brought the cavalry up in front, whilst the bulk of their infantry, in four columns, were sent to turn to the French right, and scattered over ten or twelve miles of country. This was the cause of their total defeat; for, after the four columns had been detached, they found themselves in presence of the whole French army, to which they could oppose only their right wing and the reserve. Had their columns been victorious on the left, of what use would the cavalry have been? They could never have cut in on the fugitives in time to prevent the French reaching the defiles of Bellawitz and Schlappanitz; but had the infantry been in front and defeated the French whilst, the allied cavalry gained ground to their flank, not a soul would have escaped from the field. The French remarks on this battle are most amusing: they say, "Les Russes avaient conçu un plan de bataille contre une armée qu'ils ne voyaient point; et de plus, ils admettaient l'hypothèse que les Français resteraient immobiles comme des termes."

Some of the Russian and the French cavalry regiments of the Guard met in this battle and fought with great bravery; the Russians were overthrown. It is said that many cavalry engagements took place on the right, where the Russians were well commanded by General Uwarow, who, with Bagration, effectually resisted the attacks of Lannes and part of the French cavalry under Murat; but of these engagements I can find no particulars.

The retreat of the allies was covered by 22 squadrons of Austrian cavalry and some regiments of Cossacks; these latter soon left the field, but the Austrian hussars of Hesse-Hombourg, Szeckler, and Oreilly remained to the last. These brave regiments, in spite of a destructive fire from the French artillery, held their ground between Telnitz and Aujezd till the whole of the infantry had passed, and nobly repulsed a brigade of French dragoons which tried to cut in on the line of retreat.

Often did the Austrian cavalry thus gallantly cover the retreat of their defeated armies, and enable them to take the field again and again.

After the battle of Eckmühl three Austrian regiments of cavalry held their ground against two French divisions. At Egolfsheim, and the following day at Ratisbonne, 40 squadrons opposed the whole French cavalry, and gave the army time to recross the Danube. For three hours they held them in check by constant and repeated charges, in which 1000 men were killed or wounded, and then effected their retreat in safety through the town by occupying the entrance with some companies of infantry. The French cavalry was commanded by Bessières, and Napoleon was himself present, being wounded in the heel at the close of the fight.

In the first campaigns of this war the Prussian cavalry did nothing worthy of its former reputation. At Jena and Auerstadt it was defeated, and followed up with such untiring perseverance by the French horse that the Prussians were effectually prevented from again assembling in the field. The French gave them no respite, and pressed on to Berlin.

The campaigns of 1806 and 1807, in Prussia and Poland, show that cavalry, though numerous and brave, can seldom effect anything great when the armies to which they are attached are acting on the defensive.

When the General's object is to maintain himself in the field only, he tales up positions from which it is difficult to dislodge him, but in which the cavalry can act only partially in repelling the attacks of the enemy.

Beningsen, though forced by the plan of the campaign to act on the defensive, made good use of his cavalry at Pultusk: with them he masked the position of his army, and began the battle advantageously on his side; but his example would be a dangerous one to follow in the face of an enemy superior in that arm, for, if they succeeded in defeating the horse in front and followed them into the position, the destruction of the whole army might ensue.

The Russians were in position between the roads leading from Pultusk to Sierock and Nowemiastow. They had Pultusk with the Narew on their left, the road to Ostrolenka in the rear.

Beningsen only wished to engage the enemy partially, to retard his advance, in order to gain time to fall back; but the ground was heavy for the artillery, and the advance of the French so quick that he was obliged to engage his whole corps to maintain his ground.

He formed his army, consisting of 66 battalions and 95 squadrons (about 45,000 men), in two lines, between Moczin and Pultusk. In front of the left wing 10 battalions and 20 squadrons were pushed forward under Bagawort, and 12 battalions and 10 squadrons in like manner in front of the right wing under Barclay de Tolly.*

Two thousand yards in front of his line of battle he distributed the cavalry by regiments, en échiquier, with wide intervals, and 500 yards in front of them again a line of Cossacks, who stopped the enemy's skirmishers and obliged the French to deploy their advanced guards to drive them back.

Lannes advanced in six columns against the Russian position. The numbers of the French are differently estimated, but there can be no doubt that they were more numerous than their opponents.

The Russian cavalry of the left wing drove in the French cavalry which preceded the columns on that flank, and charged the infantry, some battalions of which were ridden over: however, their success was only partial; for, though they retarded the advance of these columns of attack, they could not stop them.

In the centre General Dorochow retired slowly before the enemy, till their columns were within range of the Russian batteries; then the curtain was rent, the Russian cavalry withdrew, and the French found themselves exposed to a murderous fire.

*Hist. de la Cavalerie, par L. A. Unger.

On the right Barclay de Tolly had been attacked, but had repulsed the French cavalry.

The French had now brought up their artillery and the remainder of their forces; Beningsen withdrew the cavalry behind the infantry.

The Russian advanced corps on the flanks were driven in by the French with the loss of some guns which they afterwards recovered. The flanks were reinforced with cavalry and infantry, and the Russian army formed in three lines: the first deployed, the second in columns, the third composed of cavalry.

The left wing, supported by 20 squadrons, then advanced, charged, and drove the French back. On the two sides upwards of 3000 men were killed and wounded, and the Russians made 700 prisoners.

The battle of Eylau, which succeeded in the campaign, was remarkable for the grand charge of cavalry under Murat, who led 72 squadrons forward against the Russian position; but this I shall have occasion to speak of hereafter.

After Eylau a cavalry engagement of some consequence took place; the French having upwards of 40 squadrons in the field. They were defeated and driven across the Trischinz in disorder by the Russian horse.

On the 5th, 6th, and 7th of June the allies had good opportunities of using their numerous cavalry to advantage, but they failed to avail themselves of them.

On the 10th of June, 1807, at the battle of Heilsberg, the allies had 205 squadrons, of which 27 squadrons were Prussian, the remainder Russian. With such a numerous body of horse at their command, the army took up a defensive position; strong certainly, but one in which their superiority in cavalry proved of no avail.

The Russian horse were in reserve the greater part of the day. The Prussians behaved well, and executed some gallant charges; but the battle was undecided. The allies remained in position during the 11th, and in the evening retired towards Bartenstein.

Then succeeded the battle of Friedland, where the allies exposed their cavalry in line to the destructive fire of the French batteries during the greater part of the day: and, when they had been well shaken and pounded, the French cavalry charged and defeated them.

In these campaigns the constant system of defensive warfare, the making cavalry replace infantry in battle, and uselessly exposing them to the fire of artillery, so demoralised and discouraged that of the allies, that in 1812 they could hardly be brought to face the French horse.

In 1813, at Lützen, the very favourable ground fur the action of cavalry gave Napoleon the impression that the allies were at last about to avail themselves to some purpose of the 18,000 horsemen they had in the field. To these Napoleon could oppose only 5000 *mounted soldiers*: for his old cavalry, inured to war and accustomed to victory, had been entirely destroyed the preceding year in Russia; and all his power, resources, and genius had been exerted in vain to replace them; and so conscious was he of the danger of meeting the numerous cavalry of the enemy in the open field, that, on hearing that Ney had been attacked, he moved up the troops to his assistance, formed in large squares of several regiments of infantry, with artillery on their flanks and cavalry in the rear.

When the battle began Ney was alone on the ground: he threw his infantry into the villages, and maintained himself there in spite of all the efforts made to dislodge him. The allies grew obstinate and persevered, feeding the fight with more and more infantry, till their place in line was obliged to be filled up by regiments of cavalry, which were apparently forgotten whilst this contest on foot was carried on. During this time Napoleon brought the whole of his army into line, and his artillery made great havoc amongst the imposing array of German and Russian cavalry with which the background of the picture was filled up.

The day closed in; the troops were tired out; all the infantry, with the exception of the Russian Guard, had been engaged, whilst the French had still fresh troops at hand for any emergency.

The allies had a reserve of 50 squadrons of Russian horse, but the opportunity for using them had passed away: they could do nothing against the French, who were established in broken ground, with a battery of 60 guns in position.

According to Frederick the Great, villages should be occupied only for defensive purposes. An army acting on the offensive should lose no time in contesting their possession, if they can gain their point by turning the enemy's position.

Had the allies done this, and pushed forward their cavalry to attack the French corps in the plain whilst coming up to Ney's assistance, the result might have been widely different; and had they been successful, Ney's corps must have laid down its arms.

That the French infantry of that year were incapable of meeting the German cavalry successfully in the open field, was proved by Colonel Dolfs, shortly afterwards, at Haynau, where, at the head of 20 squadrons of Prussian cavalry, he attacked General Maison's division, formed in 8 squares and protected by 18 guns, and in less than 15 minutes swept over them, killing, wounding, or making prisoners the whole force and capturing their guns. The only men that escaped were a detachment of French cavalry who took care to get a good start.

With all this, when the French were defeated at Leipzig, they made good their retreat, though the allies had 80,000 cavalry in the field! This is not to be satisfactorily accounted for by the intervention of a petty river and the blowing up of a bridge by the French.

In 1814, when the allies, fighting on the soil of France, defeated Napoleon at La Rothière, they allowed his army to escape again, although they were themselves strong in horse.

In 1815, after the defeat at Waterloo, though the Prussian cavalry started in pursuit or the French from the field, the French soldiery not only escaped in large numbers, and rallied beyond Paris, but their cavalry, at Versailles, defeated and made prisoners an entire brigade of Prussian horse.

Cavalry is often doomed to total inaction in battle from the manner in which armies are formed for the fight. Generals attempt to use armies as they would machinery, quite forgetting that such different component parts as infantry, cavalry, and artillery cannot always work together. A happy combination of the three, or a grand coup struck by cavalry alone, is rather a rarity in modern warfare. Isolated gallant charges of cavalry are heard of (as, for example, the charge of the 3rd Dragoons at Moodkee, and that of Major Unetts's squadron at Chillianwallah); but such charges, though executed with the greatest energy, never go beyond the limits of what is expected of those detachments of cavalry which are attached to each division of the army in the field. Thus we see a display of gallantry, and even of skill, without any grand result.

The fact is, modern tactics hold the cavalry in leading-strings. The system is a timid one, and made up of "ifs" and "buts," *words* which ought to be unknown to cavalry soldiers.

Instead of trying at once to strike home when the opportunity offers—instead of pouring the whole cavalry of the army on the enemy's flanks or rear, they fritter away their strength, march and countermarch, advance perhaps one-third of their force against the enemy, keeping two-thirds in reserve to guard against unknown dangers. They attack under cover of batteries established to insure them against the ill consequences of failure, etc. In short, examples of successful cavalry actions (we mean simply and purely cavalry actions) are to be found only in few and isolated instances, where the horsemen, acting on the spur of the moment, and forgetting their tactics altogether, were led away by the bold example of some chivalrous leader. Such was the case when Murat attacked the Austrians near Dresden, leading part of his cavalry round their left flanks whilst the remainder attacked in front. Murat fell upon them from both sides at once, and, sword in hand, captured sixteen guns and made 15,000 prisoners. This was, assuredly, one of the most brilliant cavalry actions of modern times.

A curious circumstance, and one worthy of remark, is, that in the late wars, wherever the cavalry were made use of in large bodies, the greatest confusion generally ensued: for instance, at Craonne the Russian cavalry, though successful in their first charge, got into such disorder, and so mixed up together, that the whole mass was withdrawn in confusion from the field, to save it from destruction. Luckily for them the French cavalry, under Nansouty, were far away on the right flank when this happened, or they would hardly have effected their retreat in safety.

This confusion in action may be partly accounted for by the depth of the formations, the number of lines ranged one behind the other when going into action, the uniformity in the dress of the different regiments engaged, and the large squadrons which, once dispersed, can rally only with the greatest difficulty, and always require a long time to do so.

Lines of cavalry following each other must get into disorder if any part of them be driven in by the enemy, for there are no outlets for the fugitives.

The cavalry of Frederick the Great generally doubled the extent of their intervals when advancing; their second line was only partly deployed: it followed in échellon, or with intervals the breadth of a squadron.

The infantry of those days formed in line, and brought every musket into play against cavalry charging them.

We English pride ourselves on having adhered to this system: and after the French had, in columns and masses, walked over every army in Europe, they were defeated by our lines of infantry in the Peninsula. The French cavalry, led on by Lasalle, Montbrun, Latour-Maubourg, Valmy,—all cavalry generals of high renown,—drew back and shrunk from their fire; and the British foot-soldier remained unconquered, and gained many a laurel wreath, though he did stand and fight in line!

Yet neither lines nor masses of infantry stopped Seidlitz and his glorious horsemen. At Kunersdorf and Zorndorf there stood opposed to him, for every yard of ground, ten or twelve infantry soldiers.

The want of success on the part of cavalry in later years cannot, therefore, with justice be attributed to the different formations of the infantry: and I would fain ask whether the ground was less good, or circumstances less favourable for the employment of cavalry, at Lützen, Dresden, Leipzig, and Craonne, than at Rossbach or at Zorndorf?

The humiliating confession is forced upon us, that, if cavalry have fallen from their high estate, they can blame only themselves and their own tactics.

3

Cavalry in General

I t must be admitted that in modern warfare we have rarely seen the happy combination of excellent cavalry commanded by a perfect cavalry officer. The passage is somewhat inflated, according to the genius of the language in which it is written; yet General Foy scarcely exaggerates the amount and union of qualities requisite to form a first-rate cavalry leader.

"Après les qualités nécessaires an commandant en chef, le talent de guerre le plus sublime est celui du général de cavalerie. Eussiez-vous un coup-d'oeil plus rapide et un éclat de determination plus soudain que le coursier emporté au galop, ce n'est rien si vous ne joignez la vigueur de la jeunesse, des bons yeux, une voix retentissante, l'adresse d'un athlète, et l'agilite d'un centaure. Avant tout, il faudra que le ciel vous ait departi avec prodigalité cette faculté précieuse qu'aueune ne remplace, et dont il est plus avare qu'on ne le croit communément, la bravoure."*

Of all arms, cavalry is the most difficult to handle in the field.

It cannot engage an enemy except where the ground is favourable.

It is always dependent on the condition of its horses.

It is easily dispersed, and it easily gets out of hand.

However brave and intrinsically good, it is of no use without good officers.

The qualities requisite in a cavalry leader are, a good eye for country, and a quick one for the enemy's movements, great energy, courageous decision, and rapid execution.

Histoire de la Guerre de la Péninsule.

No wonder, therefore, that cavalry has not always developed its power and resources in the field; for, placing all other considerations aside, how few examples does history afford of cavalry being well led and commanded! When well led it has been invariably successful.

Cavalry ought to be at once the eye, the feeler, and the feeder of an army. With good cavalry an army is in comparative security, and in a condition to march into and subsist upon an enemy's country. It reaps the fruits of victory, covers a retreat, and retrieves a disaster.

With it the effects of a defeat are not always fatal, and with it the army can again resume the offensive.

In defensive warfare it has seldom achieved great deeds, for to act a *passive part* in war is contrary to the spirit of Cavalry Tactics.

When badly organised and badly led, the more numerous it is, the more useless. [Witness the engagements of Medellin, Ciudad Real, Ocana, and Alba de Tormes, where the Spanish horse fled the field, and left their infantry to be cut down by the victorious French.] It eats up the supplies of the army, and is in battle a dangerous ally. It gets out of hand in action, and, instead of injuring the enemy, entails defeat on itself and on the army to which it belongs.

We have seen that individual prowess, skill in single combat, good horsemanship, and sharp swords, render all cavalry formidable.

That light and active horsemen have, in the long ran, prevailed over heavily-equipped cavalry, and that speed and endurance are qualities to be highly prized in the horseman.

Therefore, if these views be correct, then our European cavalry is not organised on an efficient system.

For the present riding drill makes few good horsemen.

The swords, blunted by steel scabbards, are not efficient weapons.

Speed and endurance cannot be expected from horses that are over-weighted.

Celerity and precision of movement cannot be attained with large, unwieldy squadrons.

Nor can decision be expected on the part of the leaders, under a system of "Pivot Flanks," and "Right or Left in Front."

T HERE ARE THREE KINDS OF CAVALRY now established in Europe—
Heavy Cavalry, Dragoons, and Light Cavalry.

The different size, strength, and qualities of men and horses seem to require them to be thus divided into heavy, middle, and light; for a horse physically unfit to carry a cuirassier would be lost to the service unless made use of in the dragoons or in the light cavalry. And where there exists a difficulty in finding sufficient horses for the purposes of war, a system by which the greater number of animals can be made available is the one which has been generally adopted.

The nations of the European Continent, who take the field with large armies, require a numerous cavalry: they cannot have them all good; some cannot obtain horses, others cannot afford the heavy expense, and thus they are of necessity reduced to a system of expedients. But England, rich in men, money, and, above all, in horses, should, in this particular, avoid imitating foreign armies, and, instead of reducing her cavalry to a par with those on the Continent, she ought to make her own cavalry so superior as to defy comparison and all competition.

The Heavy, Middle, and Light Cavalry have different parts assigned to them in war, not one of them being fit to perform all the duties required of horse-soldiers in the field.

HEAVY CAVALRY

Composed of large men in defensive armour, mounted on heavy, powerful horses, are held in hand for decisive charges on the day of battle, and their horses are so deficient in speed and endurance (being so overweighted), that they require light horse to follow up the enemy they have beaten. The greatest possible care is taken of this sort of cavalry in the field. They do no outpost duty, no foraging, no reconnoitring: they cannot be made use of even to escort a convoy, because, if kept out long on the road, their horses fall off in condition and become incapable of carrying their riders. They are calculated only to show an imposing front in the line of battle, and their history proves them to be more formidable in appearance than in reality.

Dragoons

Were originally intended to be a sort of hybrid corps, or infantry mounted on horses, in order that (like the Janissaries in Suwaroff's war) they might arrive with more expedition at the position in which they were to fight on foot; and in a battle they formed line and acted with the infantry. At first they were denominated Arquebusiers à Cheval; afterwards they were named Dragoons by the famous Count of Mansfeldt, in comparison with the imaginary Dragons represented as spitting fire and being swift on the wing. The Swedes first used them as light horse against the Croats, a light cavalry of the Austrian Emperor. At a much later period the English and Hanoverians mounted them on powerful horses, substituted trumpets for the drums then in use; and thus, by degrees, the dragoons took a higher place with the cavalry. Still later, they were however expected to act both as horse and foot soldiers. It was a favourite project of Napoleon thus to organize them for both services; but, after much loss of time and great expense, finding it did not answer, he took away their muskets and bayonets and gave them carbines; and they were reorganized and sufficiently well mounted to charge with advantage, being at the same time lightly equipped in order that they might be serviceable as skirmishers, foragers, etc. The difficulty of Napoleon's first intention is easily understood, if we consider the time required to form a cavalry soldier and the time required to form an infantry soldier. If we succeeded in bringing a body of men, in time of peace, thoroughly to understand the duty of both, how could we keep our regiments of dragoons complete in time of war? How could we then find time for this long double drill and training? Then, again, bring your regiment of dismounted dragoons into action, and what would follow? It would be less numerous than a real infantry corps opposed to it; the long swords and spurs of the dragoons would be in their way, particularly if skirmishing; and should a few of the enemy's light horsemen make a dash at the led horses, the dragoons would run a good chance of becoming only infantry for the remainder of the campaign. Dismounted cavalry have done good service in covering a retreat, in defending defiles and passes against cavalry, and in pushing forward to seize bridges and halting to maintain them; but they would be quite out of place if used in storming positions, or if expected to take their post in line of battle with the infantry.

LIGHT CAVALRY

The service required of these is the most important in the field. They are called upon to watch over the safety of the army, and they are constantly hovering in advance, on the flanks, and in the rear of the columns, to prevent all possibility of surprise on the part of the enemy. In enclosed countries they are supported by light infantry: in the open country the light cavalry push on and keep the enemy at a proper distance from the army; they are constantly employed in cutting off the enemy's supplies and communications, in reconnoitring, etc. This varied and often impromptu work requires a combination of numerous qualities in officers and men. And, in addition to all these duties, peculiarly their own, they often have to perform also those expected of the heavy cavalry; and with what success they have done this I shall presently endeavour to show.

4

LIGHT AND HEAVY CAVALRY

*"Armour protects the wearer, and prevents him from injuring others."**

THE POWER OF HEAVY CAVALRY lies in the strength and breeding of the horse, and the courage and activity of the rider. The size of the rider, his cuirass, defensive armour, and heavy equipments, detract from the speed and lasting qualities of the horse, and only render the man helpless; for they impede and unfit him for any exertion in which activity and endurance are necessary.

If a heavy-armed horseman gallops and exerts himself only for a few minutes, the horse is beat by the weight, and the rider is exhausted in supporting himself and his armour in the saddle; his sword-arm hangs helplessly by his side, he can hardly raise his heavy broadsword: such a man is at the mercy of any light horseman that may turn upon him.

Speed is more than weight: in proportion as you increase weight you decrease speed, and take from your cavalry that impetus which ought to be its principal element. We are not the only military nation who have committed this error. With horses far inferior to ours—inferior both in size and in breed—our neighbours have gone for weight. In the last war the French cuirassiers were reduced to charge at a trot, their horses being unable to carry such weight at a quicker pace. In their attacks on an enemy's position, the losses they sustained from the want of speed were frequently awful. Under the improved fire of the artillery and infantry of the present day, these slow attacks never could be carried out at all.

Heavy Russian cuirassiers, when opposed to the Turks, were obliged to form in close columns or in squares, requiring artillery and infantry

*This saying is attributed to one of the German emperors.

to protect them from the sharp scimitars of the Moslem. These Turks had no discipline, no lances,—had nothing but their good swords and steeds to trust to.

And what, in battle, is the real value of the cuirass or other ponderous defensive armour for the body? So long as arms, legs, and heads are unprotected, it signifies little that the chest he covered with armour, for the moment either of the horseman's arms is wounded (it signifies not which arm) he is at the mercy of his adversary.

The weight of the armour only renders it more difficult for the cuirassier to defend himself against a man who is free from encumbrances, and who, if furnished with a proper sword, can lop off a limb or kill his opponent's horse at one blow.

The nations of the Continent, as I have previously observed, cannot obtain sufficiently well-bred horses of the required size and power, and they are therefore obliged to mount their heavy dragoons on large, clumsy, and slow horses. To make the best of a bad job, when their men are thus mounted they case them in armour, in order that they may have a better chance of reaching the point of attack alive, that they may be inspired with confidence as to the protective power of their shining breastplates, and that they may work upon the nerves and imagination of the enemy by their imposing appearance as "men in armour." But the brave light horseman soon finds out that, whatever they may be to the eye, they are in action scarcely more formidable than the men in armour who ride (or used to ride) once a year in my Lord Mayor's show.

England, if reduced to mount her heavy dragoons on Barclay and Perkins's dray-horses, would most likely do the same as the French, or arm the men cap-à-pied; but, whilst no dearth of horses has yet reduced her to this expedient, it is truly pitiful that she should copy from such bad originals as the continental cuirassiers! The Prussian Major-General of hussars, Warnery, in speaking of the English cavalry, says—

"The English have everything which can be desired to form an excellent body of cavalry of all species; their light dragoons might, and do, surpass everything which we have ever seen of that nature; and, as their cavalry is not numerous, they have the greater facility in being select in its composition, both in men and in horses, without being obliged to have recourse to other nations, or to look out of their own island,—an advantage which few countries possess."

If England could mount her cavalry on horses combining more power and size than any in Europe, with more breed, speed, and activity than any now in Asia, she ought to endeavour to adopt a system which, in giving full scope to their excellence, would enable the English horse to bear down and ride over the disciplined resistance of continental troops. By taking a lesson from the Asiatics, she might so arm and instruct her dragoons as to make them equal to any of the people of the East in single combat.

Heavy cavalry should have the largest and most powerful horses, but the men and their accoutrements should be light. If you weight the powerful horses with heavy men and accoutrements, you bring them down to a level with smaller and weaker horses. Thus a great heavy man in armour, on a fine strong horse, could not catch or ride down a Cossack on a good pony; but the same horse, with a light active man on his back, would ride down a dozen of such Cossacks, one after the other.

In a charge, the same horses with light weights will, by their speed and impulsive power, ride down or over obstacles which would certainly stop them if heavily weighted. The heavier the man, the less available the high qualities of the horse, and the less formidable the man on his back.

What (except, perhaps, the want of opportunity) is to prevent our armour-clad Household cavalry from meeting with the same fate at the hands of some active and determined light horsemen, as befell the brave French cuirassiers when they were shot and speared off their horses by the Cossacks?

If English dragoons were properly organized, properly furnished with offensive weapons, and duly impressed or imbued with confidence in the strength and speed of their horses, in their own riding, and in the destroying power of their swords, no numbers could daunt them. A few such men could hang like shadows round an enemy's cavalry column, reconnoitre their movements, approach, dismount, and pick off their officers; while the enemy could neither catch them nor drive them away. Then, again, in an emergency, our Englishmen could make their way across country where no foreign dragoons could ride and follow.

Arm and form your men according to the system which I propose, and which will be detailed in another chapter, and, in my humble opinion, you go far to secure the following advantages:—

When acting in bodies, no rattling of swords' scabbards would announce their approach to the enemy, and prevent the words of command of their own officers from being heard.

No unwieldy squadrons would prevent a speedy advance over difficult ground, or exhaust the horses by pressure in the ranks, causing confusion and occasioning breaks and gaps in the line.

No cavalry could withstand the speed and power of their charges, or escape from their death-dealing blades.

No Moslem could oblige them to seek shelter behind batteries or columns of infantry.

No light cavalry could swarm round their columns, spearing and shooting the outside files.

No fire of infantry could be repeated sufficiently quick to inflict a loss of an amount to check them in their charge; for they would he upon the infantry with the speed of lightning.

Finally, commanding officers, freed from the incubus of pivot flanks, and right and left in front, would act, boldly, resolutely, quickly; and thus lead our cavalry to gallant deeds and to almost certain victory.

T HERE HAVE NOT BEEN wanting on the Continent military writers to weigh the respective value of heavy and of light cavalry, and to point out the long series of successes gained by bold riders lightly equipped, well mounted, and armed with proper weapons. It is now many years since the Prussian General, Warnery, published his interesting remarks on these subjects.

The Albanians, whom Warnery mentions in the first instance, made themselves truly formidable in the fifteenth and sixteenth centuries. They went by various names, and seem to have had little right to be, called Albanians, for their bands were composed of daring adventurers from the Morea, from ancient Thessaly and Thrace, from Servia, from Dalmatia, and other regions, as well as from Albania. The contemporary Italian historians generally call them Stradiotti. Apparently, they brought their horses as well as their arms with them into Italy, whence their renown as an indefatigable light cavalry was spread over Europe. They had, no doubt, taken lessons of the irregular Turkish cavalry.

"These Albanians," says General Warnery, "served in the field exactly as the hussars of our own times; and if they had the good fortune to throw the powerful gensdarmes into a little confusion, they soon made a great carnage amongst them; for, being hand to hand, and pell-mell with them, those heavy horsemen could make no use of their lances, and, in fact, could scarcely move themselves. One fact occurs in the military history of the period, which appears almost incredible, but is, nevertheless, true and certain: the Swiss foot, armed simply with pikes and halberds, attacked and defeated the gendarmerie in the plains, and particularly at Novara, where the heavy French gensdarmes were almost annihilated.

"Philip de Comines mentions, that in his time, when the French and Venetians blockaded Verona, defended by the troops of the Emperor Maximilian, a party of Albanians sallied from the place, and skirmished with the French gensdarmes, and that each Albanian took one of the gensdarmes prisoner, and led him into the town in triumph. [At present, however, this would not appear so very extraordinary, as a single Russian, or Cossack, has frequently taken two cuirassiers in one day.]

"In a march of ten German miles, supposing it to be commenced with equal numbers, the hussars would certainly have the advantage over cuirassiers. In the open country they would very much harass and dishearten heavy cavalry by continual skirmishing and hanging upon their flanks and rear; and the effect would be very much increased if the heavy horse should be provoked to charge, even though they should be so fortunate as not to be in disorder after charging.

"But in such a length of march there must at last be some *défilée*, or other obstacle, which would oblige this heavy cavalry, already much fatigued, to break off; and this is the moment for the light troops to act with the greatest vigour, and by continual pressing upon them in such situations (in which they can neither prevent being attacked, nor take their revenge), they will at length lose confidence; and the instant they either charge or disperse, they are generally certain of being vanquished.

"To remedy this disadvantage, the King of Prussia directed all his cuirassiers to be practised to the hussar exercises, which was, certainly, so far useful; but their horses are not proper for such light and active service.

"Seidlitz, whose regiment ought (for the useful) to serve as a model for all the cavalry in the universe, allowed that, in a march of length, he should not be able, with his whole regiment, to resist 600 good hussars.

"All heavy cavalry who lose confidence, or disperse, in presence of light, are lost: if they determine, by one great effort, to extricate themselves (at least for some time), the light retreat swiftly, *à la débandade*, in all directions.

"General Werner, with 700 hussars, completely destroyed the dragoons of the Archduke Joseph,* afterwards Emperor, by skirmishing, harassing, and hanging upon their flanks and rear in the manner above described. Those dragoons were commanded by General Caramelli.

"The Prussian hussars are equally capable of every nature of service. In regular battles they have rendered the service of cuirassiers; they never hesitated to attack in close squadron whatever they have met with, which was never known before them to have been done by the hussars of any other nation; it being the general opinion, and even of the hussars themselves in those services, that the nature of that arm is not proper to act in line, nor do they scarcely ever make their appearance during an action, which could originate only in the ancient prejudice that the goodness of cavalry consisted exclusively in the 'height of the man and horse.'"†

Yet the gallant cavaliers of Gustavus Adolphus and Charles XII were never mounted upon any other than Swedish, Friesland, or Livonian nags or ponies,—and nevertheless what prodigies did they not perform?

Colonel Marainville, the French military commissioner in the Austrian camp, speaking of the Prussian hussars, says,—"Le jour de la bataille du 5ème Décembre, j'ai vu de ces hussards pousser une grande garde de cavalerie (cuirassiers) jusque dans le village de Leuthen, où

---

*An Austrian dragoon regiment has 1400 horses, and upwards, in time of war.

†Major-General Warnery, *Remarks upon Cavalry Tactics*.

nous avions de l'infanterie, et un de ces hussards fendre la tête a un cuirassier à trente pas de la première maison de ce village."*

Even in olden times, speed not weight was regarded as the first quality in cavalry. Marshal Saxe said,—"Cavalry which cannot charge at speed over a couple of thousand yards, to pounce upon the foe, is good for nothing in the field." All the tendency of modern times and modern experience is to impress upon us the paramount necessity of speed. Even of the infantry Napoleon was accustomed to say, "Arms win fewer battles than legs."

The great improvement made in fire-arms, and the increased range of the infantry musket, leave but little chance for cavalry, unless the speed with which they can pounce upon the infantry lessens the number and the effect of the discharges to be received during their advance. How can this be done with cuirassiers? Ere they could close upon the foe, if saved themselves by their armour, most of their horses would be wounded or killed, and where is then the advantage of a cuirass?

I have been assured that Captain Minié, with the "culot" ball, hit a mark seven times out of ten shots at a distance of 1800 yards, and even at that distance it is supposed that the ball "primative " would take effect through the cuirasses.

At some of the experiments made in England, a Minié ball passed through an earthen breastwork three feet thick, and killed a soldier standing behind it, smashing his skull to pieces! *What sort of cuirass would resist such a bullet?*

In a *mêlée* the cuirass may save the man from a sword-cut or point in the chest, but he is only the more vulnerable about the arms and legs; and when either bridle or sword-arm is injured the cavalry soldier is at the mercy of his opponent.

As fair a test of the relative efficiency of men in armour and those without is to be found in the charges and conflicts of Cavalry at Waterloo. Our men had no armour, they were overmatched greatly in numbers, yet when they charged they drove the cuirassiers before them; and as for single combatants, if we take the life-guardsman Shaw, we

*Extract from a work by Von Stuhr.

have a fair proof of the superiority of the man unincumbered with armour: it is said he killed several of his steel-clad opponents in fair fight, and when set upon by four of them at once, he killed three, and was then disabled by a pistol-shot from the fourth.

Captain Siborne thus describes a charge of cavalry at Waterloo:—

"They are the far-famed cuirassiers of France, led on by Kellermann: gallant spirits that have hitherto overcome the finest troops that could be brought against them, and have grown grey in glory. Trumpets sound the charge; in the next instant your ears catch the low thundering noises of their horses' hoofs and your breathless excitement is wound to the highest pitch, as the adverse lines dash together with a shock which at the moment you expect must end in their mutual annihilation. Observe the British, how they seem to doubt for a second in what manner to deal with their opponents. Now they urge their powerful steeds into the intervals between the necks of those of the cuirassiers. Swords, brandished high in air, gleam fitfully in rapid succession throughout the lines; here clashing together, there clanging against helmet and cuirass, which ring under their redoubted strokes. See, the struggle is but a moment doubtful: the cuirassiers, *seemingly* encumbered by their coats of mail, are yielding to superior strength, dexterity, and bravery combined; men and horses reel and stagger to the earth; gaps open out in their line; numbers are backing out, others are fairly turning round; their whole line now turns and breaks asunder into fragments: in the next moment they appear, as if by a miracle, to be swept off the crest of the position, and being closely and hotly pursued by the victors, the whole, rushing down the other side of the ridge, are snatched from your view."

Sergeant-Major Cotton relates the following encounter:—

"A hussar and a cuirassier had got entangled in the mêlée, and met in the plain in full view of our line; the hussar was without a cap and bleeding from a wound in the head, but that did not hinder him from attacking his steel-clad adversary. He soon proved that the strength of cavalry consists in good horsemanship and the skilful use of the sword, and not in being clad in heavy defensive armour.

"The superiority of the hussar was visible the moment the swords crossed; after a few wheels a tremendous facer made the Frenchman reel in the saddle, and all his attempts to escape his more active foe became unavailing; a second blow stretched him on the ground, amidst the cheers of the light horseman's comrades, the 3rd German hussars, who were ardent spectators of the combat."*

Captain Ganzauge, of the Prussian lancer guard, in his *Kriegswissenschäftlichen Analecten*, gives various instances in the campaign of 1813 of actions fought between the Cossacks and French cavalry; all of them most interesting, for they show how cavalry, by its equipment and system of tactics, can be made so helpless as to be unable to dispose of such despised enemies as the Cossacks. As the work is not generally known in this country, I will proceed to give a few brief extracts, which I have translated from the original.

"On the 19th of August, 1813, when the armistice had expired, the French troops began to push the allies back on Berlin and Potsdam. Colonel Bichalow received orders to make a recognisance in the direction of the Lüchenwalde with a regiment of Don Cossacks. These men had bivouacked on the Treboin road, and advanced through Scharfenbrück and Waltesdorff. The French picquets retired at our approach, and afforded us a full view of the fields to the north and east of the Lückenwalde. Presently a large body of cavalry issued in haste from the town, and formed in our front in close column of squadrons; their skirmishers fell in on their flanks, and we had this heavy column only before us. The Cossacks could gain but little against so large a force; but as there was no great risk in assailing it, they were ordered forward. The French advanced at a trot, and, to prevent the. Russians getting in betwixt the squadrons, they closed up and bore right down on the centre of our line, which naturally opened out; the Cossacks attacking the flanks and rear of the column. The French, having no one in front to oppose them, halted, whilst their tormentors kept spearing the flank files and firing into the mass, which soon got into complete confusion and could undertake no evolution of any sort. The Cossacks, though they never attempted to disperse the mass by a dash at them, still, conscious of their superiority in riding, continued to shoot and spear them, executing partial charges when opportunity offered.

*Cotton, *A Voice from Waterloo*.

Meanwhile the flank files of the French faced outwards and unslung their carabines, and, thus formed in square, they kept up an irregular fusillade for about half an hour. The heads of French infantry columns were now seen advancing from Lüchenwalde towards the scene of strife; and as soon as their artillery opened fire, the column of *heavy* cavalry was released from its dangerous situation. Colonel Bichalow withdrew his troops by way of Scharfenbrück, without being pursued by a single Frenchman.

"Soon after the battle of Dennewitz, the Cossack regiments, named before, were in the neighborhood of Königsbrück and Dresden. Colonel Bichalow was ordered to watch the French cavalry, which had been pushed forward towards Grossen-Hayn, and to attack them if possible. On the 18th of September we proceeded accordingly towards Esteleverda. Here we were told that the French cavalry occupied the villages to the south of Mühlberg, and resolved to beat up their quarters at once. Before we reached the heath extending between Mühlberg and Spannberg, General Slowaisky joined and assumed the command of the Cossack force, which, together with the regiment he brought up, amounted in all to 1200 men.

"I am not aware whether this meeting was the effect of design or chance. When we cleared the wood of Mühlberg, we saw the French cavalry, near Borack, partly formed, whilst parties were trotting up from the different villages to join them. The prisoners afterwards told us that their force on the ground was 2000 men.

"The French completed their movements whilst the Cossacks were forming up; they were formed in one line, *en muraille*, with a small reserve in rear.

"The Cossacks fell on, and were received with a discharge of carabines; the French did not draw swords. Their fire, at first, sent the Russians to the right about; and, whilst they were re-forming, the enemy wheeled into column and opened out, so as to get their intervals wheeled again into line. We expected they were about to charge, but their object appears simply to have been to extend their line, and prevent their being outflanked; a common mode of attack with the Cossacks.

"The arrangements being completed on both sides about the same time, the Cossacks were strictly admonished not to shrink from the enemy's fire, their officers receiving orders to cut down the first man that turned.

"Several squadrons were told off to attack the enemy in flank and rear during the conflict. All these orders were steadily obeyed; they pressed in upon the French, and surrounded their squadrons: here I saw, myself, many of the French dragoons cut down or speared after firing off their carabines, before they could draw their swords. The French steadily defended themselves at first, as well as cavalry standing still can do, against such active adversaries, who swarmed about them on all sides; however, presently, some of them turned, and their example was soon followed by the remaining squadrons. The reserve, instead of advancing to restore the fight, joined in the flight; in a short time every one was galloping towards Jacobsthal, and the entire plain was covered with scattered horsemen. Not one troop was to be seen in close order; it was a regular hunt; and most of those who were taken prisoners in it had previously fallen off their horses. At last we came upon a line of cuirassiers, in emerging from the wood, and their steady and imposing attitude brought us suddenly to a halt, without any word being given. We were quite satisfied with—our victory, and turned back to Mühlberg."*

Many more striking examples of the superiority of the Cossacks as cavalry are upon record, and ought not be overlooked or forgotten. I will quote a few from a French cavalry officer:—

"The Cossacks," says General de Brack, "were an arm which rendered the war highly dangerous, especially to such of our officers as were intrusted with making reconnoissances. Many among them, and especially of the general staff, selected by the Major-General, preferred forwarding the reports which they received from the peasantry, to going to a distance and exposing themselves to the attacks of the Cossacks. The Emperor, then, could no longer know the state of affairs."

Thus, behold even French officers not daring to expose themselves! Behold the genius of Napoleon paralyzed by the *activity* of these semibarbarous horsemen! Does not this single fact carry great weight with it?

*Captain Ganzauge, *Kriegswissenchäftlichen Analecten.*

Again, General Morand, another French officer, says, "But these rude horsemen are ignorant of our *divisions*, of our *regular alignments*, of all that *order* which we so *overweeningly* estimate. Their custom is to keep their horse close between their legs; their feet rest in broad stirrups, which support them when they use their arms. They spring from a state of rest to the full gallop, and at that gallop they make a dead halt: their horses second their skill, and seem only part of themselves; these men are always on the alert, they move with extraordinary rapidity, have few wants, and are full of warlike ardour. What a magnificent spectacle was that of the French cavalry flashing in gold and steel under the rays of a June sun, extending its lines upon the flanks of the hills of the Niemen, and burning with eagerness and courage! What bitter reflections are those of the ineffectual manoeuvres which exhausted it against the Cossacks, those irregular forces until then so despised, but which did more for Russia than all the regular armies of that empire! Every day they were to be seen on the horizon, extended over an immense line, whilst their daring flankers came and braved us even in our ranks. We formed and marched against this line, which, the moment we reached it, vanished, and the horizon no longer showed anything but birchtrees and pines; but an hour afterwards, whilst our horses were feeding, the attack was resumed, and a black line again presented itself; the same manoeuvres were resumed, which were followed by the same result. It was thus that the finest and bravest cavalry* exhausted and wasted itself against men whom it deemed unworthy of its valour, and who, nevertheless, were sufficient to save the empire, of which they are the real support and sole deliverers. To put the climax to our affliction, must be added that our cavalry was more numerous than the Cossacks; that it was supported by an artillery, the lightest, the bravest, the most formidable, that ever was mowed down by death! It must further be stated that its commandant, the admired of heroes, took the precaution of having himself supported in every manoeuvre by the most intrepid infantry; and, nevertheless, the Cossacks returned covered with spoils and glory to the fertile banks of the Danaetz, whilst the soil of Russia

*But, alas! so unwieldy, so encumbered, and, therefore, so useless.

was strewn with the carcases and arms of our warriors, so bold, so unflinching, so devoted to the glory of our country."*

Now just contrast this account of the Cossacks—full of generous admissions, particularly for a Frenchman with the wretched deeds of the regular cavalry of the Russian army, and then who will venture to assert that the organisation, the heavy, unwieldy squadrons, the puzzling tactics of the regulars, have not much to answer for?

If Cossacks, mounted on ponies, and wretchedly armed, could thus master the French regulars, in spite of their artillery, what might not be expected from them if they were mounted on well-bred, powerful horses, and furnished with really good weapons? In that war their lances were notoriously bad: so much so that there were French soldiers who received as many as twenty lance-wounds without being killed or seriously injured.

The same contrast is presented by our native irregular and our native regular cavalry in India. The first, acting on usage and instinct, and armed and mounted in their own Oriental way, are nearly always effective in the battle, or the skirmish, or the reconnoissance; the second, cramped by our rules and regulations, and, as it were, de-naturalised, are rarely of any service whatever. For a long series of years the only native cavalry we kept in India was the irregular. These corps were formed before our infantry sepoys, and many and most important were the services they rendered to us. They were always active—always rapid. The names of some of the most distinguished leaders of this brilliant light cavalry are still revered in India. If their corps had been Europeanised, and turned into regulars, assuredly we never should have heard of them as heroes.

Much more recent instances of the superiority of light over heavy are to be found.

In the Magyar war of 1848 and 1849 the Hungarians had nothing but hussars to oppose to the Imperialists' heavy cavalry and cuirassiers, and, though inferior in number, they always met the cuirassiers gallantly, and often defeated them with great loss. Indeed, in one instance, a single

*As quoted in the work of General de Brack.

squadron of the tenth hussars (Frederick Wilhelm), at the Battle of Tétény (3rd January, 1849), charged several squadrons of the Imperial cuirassiers, and defeated them. These heavy horsemen had a few days previously made a gallant charge, under Colonel Ottinger, against the Hungarian infantry, breaking two squares, and capturing the guns which flanked them; thus there could be no want of proper courage on the part of the heavies, and their defeat must be attributed to the cumbersome equipments, and the heavy, unwieldy horses.

The seventh division of Görgey's corps was stationed at Parendorf for the purpose of covering the frontier. A few miles off, at Woolfsthal, in the Austrian territory, stood Jellachich, his outposts held by the Walmoden cuirassiers. The Hungarian outposts were held by the Nikolaus hussars. Some squadrons of cuirassiers were pushed across the frontier by the Austrians into the plain near Parendorf. A body of hussars then first made their appearance, advancing at a trot, and gradually increasing their speed as they neared the enemy; and, though but a handful compared to their opponents, they rushed upon them with such speed, and in such compact order, that they broke through the cuirassiers, and scattered them over the plain, where they fell an easy prey to their more active pursuers.

On other occasions the gallantry and success of the lights were equally conspicuous. I translate what immediately follows from Georg Klaptka's 'National Krieg in Ungarn und Siebenbürgen:'—

"On the 18th of December, 1848, the enemy's cavalry, led by the *Banus*, attacked the Hungarian rear-guard near Altenburg. The enemy came from Sommerin, formed up two regiments, and opened fire with the artillery. The more numerous Hungarian artillery soon had the best of it, the enemy's line became unsteady, and showed symptoms of turning. At this moment Major Cornel Görgey brought up four squadrons of the tenth hussars from the second line, and charged the right wing of the enemy, inflicting on him a heavy loss in killed and wounded, and driving him in disorder from the field. The enemy galloped towards the reserves under Lichtenstein, which were advancing to the support, and, wrapped in clouds of dust, the defeated horsemen fled across the plain far more rapidly than they had advanced. This was the first time large bodies of cavalry bad met during the campaign; and here, as in numerous skirmishes which had taken place previously, the active hussar proved himself more than a match for the steel-clad horseman and heavy-armed dragoon of the Austrians.

"At 3 P.M. on the 3rd of January our left wing was attacked in front of Tétény.

"Our outposts had been driven in at Hamsabég at mid-day, and now the enemy pushed forward against Tétény. Zichy's brigade, about 3000 men, took up a position on the south-west of the place, across the high road, with the right resting on the heights, the left on the Danube, the centre on the wood along the banks of the river. The enemy, after a few long shots, presuming on their success at Móor, sent several squadrons of cuirassiers forward to attack the Hungarian centre. One squadron of the tenth Wilhelm hussars, which was posted there, advanced resolutely to meet them, and, led on by their officers, charged and burst into the midst of their mail-clad antagonists; a bloody and desperate *mêlée* ensued, in which great part of the cuirassiers were cut down or made prisoners; the remainder sought safety in flight"

On the 28th of February, 1849, when the Magyars were reduced to an almost hopeless condition, there was another brilliant light cavalry affair at Mezökövest:—

"On our retreat from Kerecsend and Macklár, the enemy sent a regiment of cuirassiers, with a brigade of guns, in pursuit. They attacked and drove in the rear-guard, about 2000 yards from our camp. So daring a deed within sight of our men was not allowed to go unpunished.

"The men of the 9th Nikolaus hussars sprang on their horses and galloped to the rescue. A splendid sight it was to see this swarm of light horsemen dashing in on the heavy cuirassiers, bursting their ranks asunder, cutting down, destroying, and scattering them in all directions.

"The hussars captured the whole of the enemy's guns, which, with a number of prisoners, they brought triumphantly into camp.

"The enemy's reserves formed on the heights opposite the camp, but did not attempt to recapture the guns lost by their cuirassiers."

These accounts I have translated from Klaptka, who was thoroughly a Magyar, but, as I have had concurrent testimony from others who were not of that party, I cannot suspect them of much exaggeration. But let me now add a few extracts out of a letter received from an old cavalry officer and aide-de-camp, who served on the side of the Imperialists, and against the Hungarians, or Magyars.

"May, 23rd, 1852.
"My Dear Nolan,

"I will try and answer your questions about our cavalry, and the effect of our cuirasses.

"From what the cuirassiers *say*, their cuirass saved them from many a bullet, and many a thrust, in the mêlée. This may be true, and the advantage of the armour probably is, that those who wear it fancy themselves safer, and are, therefore, morally stronger, and more ready to look danger in the face. Of other advantages of heavy cavalry over light we found now during the Hungarian campaigns: we were not in a position to employ heavy cavalry as it should be employed to reap advantage from it, and this for many reasons.

"We had but few regiments of light horse at out command; the heavies had to do outpost work, skirmishing, reconnoitring, etc., and their horses were knocked up with the weight they had to carry.

"As to the result of the engagements between them and the Hungarian hussars, I must first tell you what my opinion is in general with regard to charges of cavalry, and this opinion I formed upon the experience gained during the war.

"The success of a cavalry attack depends not so much on the description of cavalry or horse employed, as on the *determination* of the men;

"On their being accustomed to victory;

"On confidence in their leader;

"And last, not *least*, on the charge being made at the right moment.

"Thus, in the first half of the Hungarian war, the depressing moral consciousness of having abandoned their colours, together with being badly commanded, greatly influenced the behaviour of the Hungarian regiments; and, after their defeat at Schwechat, the only difficulties we experienced during our advance to Pesth were occasioned by the cold. and not by the enemy.

"At Babolna they tried to make a stand; one of their regiments formed a square, but was at once ridden over and destroyed by two squadrons of Walmoden cuirassiers, who advanced to the charge without the assistance of artillery; and this, as far as I remember, was the only instance on *our side* in which cavalry broke a square without first bringing artillery into play.

"The squadron of the 10th hussars, that did such good service at Tétény, was led by Meezey, a subaltern in the squadron at that time; he afterwards became their colonel, and his regiment proved itself the best on the Hungarian side.

"Later in the campaign of 1849 (except towards its close) the Hungarians received great reinforcements in troops, their moral courage rose, and then came the tug of war.

"New regiments were raised, and in numbers they were superior to us; but, of course, it was only the old hussar regiments that were formidable: indeed they behaved most gallantly, and on all occasions charged home at our cuirassiers and heavy dragoons, but they never liked to close with our Polish lancers.

"I quite agree with you that the strength and lasting qualities of the horse make the cavalry soldier formidable, and that, therefore, the animal should not be made to carry any unnecessary weight, which must always diminish and detract from those qualities in a greater or less degree: this is not my opinion alone, but the prevailing one in this service.

"The Hungarian saddle has now been adopted for the whole of our cavalry.

"In the bridling, saddling, packing, &c., many improvements have been made, as well as in the arming of the men.

"Our artillery is much altered for the better. In Vienna you will see several schools of equitation; and also a ' squadron of instruction,' composed of detachments from all the regiments in the service.

"Yours ever sincerely,

———————."

General Sir Charles Shaw gave the following most interesting account of the Circassian horsemen in a letter published in November 1853 in the *Morning Chronicle*. His opinions on the subject now under discussion have a peculiar value from his great personal experience of war, and his practical knowledge of military matters:—

"The noble Circassians who have been fighting against Russia, independent of Turkey, have been within this short time taken into the Turkish service; and it may be interesting to give a description, by a Prussian officer, of the Circassian cavalry, who are about to take a prominent part in the coming conflict. He says, 'The Circassian wears a pointed steel helmet, with a long horsetail pendent from it; a net of steel-work hangs down from the lower part of the helmet, protects the front and nape of the neck, and is looped together under the chin, underneath a short red vest, cut in the Polish fashion.

"He is clad in a species of coat-of-mail, consisting of small bright rings of steel intervened; his arms, from the wrist to the elbow, and his legs, from the foot of the shinbone to the knee, are guarded by thin plates of steel; he also wears close pantaloons and laced boots. Two long Turkish pistols, as well as a poniard, are stuck into his girdle. He has a leather strap with a noose, like a Mexican lasso, hanging at his side, which he throws with great dexterity over the head of his enemy; a Turkish sabre and a long Turkish musket are slung behind his back, and two cartridge-holders across his breast

"The skill with which the Circassians use their weapons is really beyond belief. I have seen them repeatedly fire at a piece of card lying on the ground, at full speed, without ever missing.

"They will pick up a piece of money from the ground while executing a charge, by bending themselves round below the horse's belly, and, after seizing the piece, suddenly throw themselves back into the saddle. They form the choicest body of cavalry in the Turkish service, and I have watched them, when charging, attack their opponents with a sabre in each hand, managing their reins with their mouths; they will spring out, of their saddles, take aim and fire from behind their horses, then jump into their saddles again, wheel round, and reload their guns as they retreat in full career. They are perfect madmen in the attack, and few troops would withstand the utter recklessness of danger they evince.

"This account of the Circassian cavalry by the Prussian officer may appear incredible to our Life Guards, Blues, and Heavies; but I do not forget that, while in 1851 and 1852 I first brought before the public the power of what is now called 'the Minié rifle,' the admirers of old Brown Bess attacked it right and left, and now there are, in 1853, upwards of 40,000 of this improved musket in the British army."

5

The Organization of Cavalry

Without farther preamble, I now proceed to offer, in as short a space as may be, all that I have to propose as a New System, or as a partial improvement upon the old one, whose soundness and efficiency some of our continental neighbours have begun to doubt. I shall express my convictions with the same frankness that I have hitherto used, without implying any disrespect to those who may entertain different opinions. After long consideration of the whole subject, I honestly believe that the main principle I propose are right. Without this conviction I would not publish at all, but with it I should feel it to be a dereliction not to offer to my brother officers, and the service in general, the results of my practice and meditation. In spite, however, of my inward conviction, I may be wrong. Therefore, though speaking out freely, I would lay down nothing dogmatically. I hope to remain open to conviction, and shall certainly entertain no ill feeling against such as may differ from me in opinion. From a comparison, and even conflict of opposite opinions, the service will be sure to gain something. The most hopeless condition to which an arm, or a science, or an art can attain is that where its professors sit down with perfect self-satisfaction, under the conviction that it has reached perfection, and is susceptible of no further improvement. True also is it, that nothing in this world can remain *in statu quo*, and that whatever does not advance must retrocede. It is a law of nature.

To possess a fine cavalry the men must be good as well as the horses, and the most delicate attention must be paid to both. Everyman may be taught to ride, but it is not every man that will make a good rider. Many who might be turned into good foot-soldiers are far from being proper materials for cavalry. It may, however, be said that, generally, Englishmen have a fondness for the horse, and a natural aptitude for the saddle. Though not "Abipones," we are certainly an equestrian nation. Left to his own free natural seat, and the Englishman beats the world in a ride after the hounds and a run across country. Since the peace of 1815 this manly sport—the best of all to form bold riders—has been taken up in some of the Continental nations; but it is indigenous—national and natural—to none of them; and, in spite of the interruption of rail-roads, we may still find at some single "meet" (without even going into Leicestershire) more riders of the right sort than are to be found on the whole Continent of Europe, if you deduct the Englishmen who are there resident, and who get up the Continental hunts, steeple-chases, &c. Our very ladies would beat, on the field, all their mathematical rid-ing-masters, and take gates, fences, and ditches, from which foreign officers of hussars or their dragoon rough-riders would turn aside in dismay to look out for a break or gap.

In our selection of men for cavalry regiments we ought to have more regard to agility, alertness, and quickness of sight, than to mere size. In fact, even with our good breeds, nearly all our horses are over-weight-ed. More than half of our lights are really heavies, and would be so con-sidered in every other European army.

It is not necessary that our hussars and dragoons should be men of five feet nine inches, or even five feet seven inches; but it is quite essen-tial that they should be active, intelligent, and quick-sighted. Now, these qualities, and in combination with great physical strength, you may find in men not exceeding five feet four inches; and here, while your men are equal in value, you improve the value and efficiency of your horse, by lightening the burden on his back. The Hungarian hus-sars, who continue to be esteemed as about the best light troops in Europe, are composed of compact, well-set, little men. In one of their finest regiments the average height did not exceed five feet four inches of our measurement.

Our light cavalry, made up of big men and heavily accoutred, is an inconsistency and a contradiction. When a man, with his arms and horse-furniture, rides twenty stone (and we have seen them of that weight), is he not out of his element in any cavalry, more particularly so in a light regiment? A fine young recruit, measuring five feet eight inches or even five feet ten inches, and being aged between eighteen and twenty-one, may not weigh much more than ten stone; but take the same individual, and weigh him after seven or eight years of service and regular living,—or take and weigh him again when he is approaching the age of thirty-five: at either period you will almost invariably find him too 'heavy for a cavalry soldier. What is to be done with him? His term of service may be incomplete, or he may wish to remain in the service, although conscious that he is no longer fit to be a horseman. Could not he, and such as he, be drafted into the infantry or foot-guards, and room be thus made for a light recruit? One regiment would gain a disciplined soldier, requiring little to be taught to him, and the other would gain what it wants, light weight.

DEFECTS, CIVIL AND MILITARY, OF THE INDIAN GOVERNMENT
by Lieut.-Gen. Sir Charles James Napier, G.C.B.

Sir Charles Napier says,— "We assume as the type of the cavalry horse the charger on a Hounslow Heath parade. Well-fed, well-groomed, well-trained, he goes through a field-day without injury, although carrying more than twenty stone weight; he and his rider presenting together a kind of alderman-centaur. But if in the field, half-starved, they have, at the end of a forced march, to charge an enemy, the biped, full of fire and courage, transformed by war-work to a wiry, muscular dragoon, is able and willing; but the overloaded quadruped cannot gallop—he staggers!

"This is the picture which should regulate the dress of horsemen; bearing also in mind the wasting sun which in India enervates man and beast.

"Our poor horses, thus loaded, are expected to bound to hand and spur, while the riders wield their swords worthily. They cannot; and both man and animal appear inferior to their Indian opponents.

"The active vigour of the Eastern horseman is known to me; his impetuous speed, the sudden volts of his animal, seconding the cunning of the swordsman, as if the steed watched the edge of the weapon, is a sight to admire; but it is too much admired by men who look not to causes. The Eastern warrior's eye is quick, but not quicker than the European's; his heart is big, yet not bigger than the European's; his arm is strong, but not so strong as the European's; the slicing of his razor-like scimitar is terrible, but an English trouper's downright blow splits the skull. Why, then, does the latter fail? The light-weighted horse of the dark swordsman carries him round his foe with elastic bounds, and the strong European, unable to deal the cleaving blow, falls under the activity of an inferior adversary!

"Look at our officers, mounted or on foot! Look at the infantry British soldier with his bayonet! What chance has an Eastern against them in single combat? Neville Chamberlaine, Robert Fitzgerald, Montague McMurdo, Charles Marston, John Nixon, Francis McFarlane, and many more, have, hand to hand, slain the first-rate swordsmen of the East. Oh no! there is no falling off in British swordsmen since Richard Coeur de Lion, with seventeen knights and three hundred archers, at Jaffa, defied the whole Saracen army, and maintained his ground. Why, then, is the Englishman inferior to the Eastern horseman in India?

"1st. The Eastern man's horse is his own property. and private interest beats the commissary in feeding; the Eastern's animal feeds better than the Englishman's.

"2nd. The hardships of war are by our dressers of cavalry thought too little for the animal's strength; they add a bag with the Frenchified name of 'valise,' containing an epitome of an old-clothes shop. Notably so if the regiment be hussars, a name given to Hungarian light horsemen, remarkable for activity, and carrying no other *baggage* than a small axe and a tea-kettle to every dozen men. Our hussar's old-clothes bag contains jackets, breeches of all dimensions, drawers, snuff-boxes, stockings, pink boots, yellow boots, eau-de-Cologne, Windsor soap, brandy, satin waistcoats, cigars, kid gloves, tooth-brushes, hair-brushes, dancing spurs; and thus, a light cavalry horse carries twenty-one stone.

"Hussars our men are not; a real hussar, including his twelfth part of a kettle, does not weigh twelve stone —before he begins plundering.

"The heavy cavalry horse, strange to say, carries less than the light cavalry—only twenty stone! A British regiment of cavalry on parade is a beautiful sight; give it six months' hard work in the field, and while the horses fail the men lose confidence; the vanity of dress supersedes efficiency. Take eight or ten stone off the weight carried, and our cavalry will be the most efficient in the world."

H AVING SELECTED THE PROPER man for the cavalry, one of the first considerations is to furnish him with proper weapons. The good workman must have good tools: the tools of the horse-soldier are his arms. These ought to be of the very best quality, and of the kind best suited to his branch of service. Like the most perfect artisan, the best trooper will lose confidence in his craft if you put the wrong implements into his hands.

THE ARMING OF CAVALRY

The devices of armament have made progress in the infantry, and enormous strides in the artillery; but in the cavalry, where the subject is of vital importance, nothing has as yet been suggested likely to make it more formidable in action.

The frequent misbehaviour of the Indian regular cavalry, which is armed and equipped after our fashion, ought to have drawn attention to this matter.

Captain Thackwell, in his *History of the Second Sikh War*, says,—

"It was incontrovertibly proved at this (Rumnugger) and other subsequent actions that the troopers of the light cavalry have no confidence in their swords as effective weapons of defence. It would have been difficult to point out half-a-dozen men who had made use of their swords. On approaching the enemy they have immediate recourse to their pistols, the loading and firing of which form their sole occupation.

"That such want of confidence must very seriously impair the efficiency of regular cavalry may be easily imagined. Very few natives ever become really reconciled to the long seat and powerless bit of the European dragoons.

[The native regular cavalry are made to use English saddles, and ride with long stirrups. To change these saddles was beyond my power; but my intent was to abolish the *egregious folly* of long stirrups.— Charles Napier.]

"The usual seat of the native is short.

"It frequently happened during the campaign that some dragoons in a charge lost all control over their horses. Picture to yourself a British or Anglo-Indian trooper dashing onwards with a most uncontrollable horse, and a Goorchurra or Sikh horseman, after allowing his enemy to pass, turning quickly round to deal him an ugly wound on the back of the head."

Again, speaking of the Indian irregulars—men of the same country, but differently armed, and riding in a short good seat:—

"Captain Holmes, of the 12th Irregulars, was the admiration of the whole army on several occasions. In his skirmishes with the enemy the mettle of his men was strikingly displayed. The irregular cavalry were conspicuous in the pursuit at Goojerat, always seeking opportunities of conflict.

"Having witnessed the charge of the Scinde horse at the battle of Goojerat, against the Affghan force of Akram, I am convinced that no cavalry could have achieved the overthrow of an enemy in a more spirited or effectual manner. They had confidence in their weapons and accoutrements. . . . . .

"The 9th Irregulars, under Crawford and Chamberlaine, earned the thanks of the Commander-in-Chief by some gallant skirmishes with the Sikh Goorchurras, who were constantly prowling about in quest of unprotected camels.

"Supported by brave and skilful men, the officers of irregulars are encouraged to gratify their noble thirst for distinction. . . . . The young heroes of the irregulars, Holmes Crawford and Neville Chamberlaine, Malcolm Tait, and Christie, would rather take into action 150 of their own men than 300 troopers of any Indian light cavalry regiment."

The Sikh war showed clearly—had any proof been wanting—how useless the Indian cavalry was when organised on the English model; whilst, at the same time, brilliant proofs were given of the superiority of the irregulars, armed with sharp swords, and having a proper command over their horses.

Nothing during that campaign was more more gallant and determined than the behaviour of the Scinde horse, whereas the distinction the regulars attained was such that it is best passed over in silence. Yet the only difference between the men composing the two arms lay in their *organisation*.

The regular Indian cavalry are useless to the public service; but the men composing it behave well when they are mounted, dressed, and armed after their own fashion.

"If a soldier of undoubted courage finds himself seated in a slippery saddle with long stirrups, cramped by tight clothes, and a sword in his hand that is good for nothing, he will hesitate, nay, more, he will refuse to charge an enemy, for if he does he goes to almost certain destruction.

"A cavalry soldier should find himself strong and firm in his seat, easy in his dress, so as to have perfect freedom of action, and with a weapon in his hand *capable of cutting down an adversary at a blow*.

"There is scarcely a more pitiable spectacle in the world than a native trooper mounted on an English saddle, tightened by his dress to the stiffness of a mummy, half suffocated with a leather collar, and a regulation sword in his hand, which must always be blunted by the steel scabbard in which it is encased,

"This poor fellow, who has the utmost difficulty in sticking to his saddle and preserving his stirrups, whose body and arms are rendered useless by a tight dragoon dress, and whose sword would scarcely cut a turnip in two, is ordered to charge the enemy: and if he fails to do what few men in the world would do in his place, courts of inquiry are held, regiments disbanded, and their cowardice is commented upon in terms of astonishment and bitterest reproach. This is truly ridiculous: the *system* and not the *men* is to be blamed."*

Now if this system, which has had a fair trial, has been found so bad in the East, why should it be supposed to be excellent when applied to our own dragoons? The colour of the men cannot make the system: the innate courage of the British soldier carries him into the midst of the enemy, not confidence in the power of the weapon he wields; for, when he has got amongst the enemy he can do no execution, —partly because he has no command over his horse and partly because his sword is not sharp enough to penetrate.

*From a letter published in the *Delhi Gazette*.

If a native horseman should not be put in a helpless seat with long stirrups, and should not be tightened by his dress, or suffocated by a leather stock; if it is necessary for him to have a sword that will cut down an enemy at a blow,—are these things less necessary to the English dragoon? or, if not quite so necessary, would they not add greatly to his efficiency in the field?

It doubtless requires great liberality and freedom from prejudice or preconceived opinion to admit that a system, on which the talent and experience of practical men has been exhausted for ages, can be a bad one.

Yet experience for many many years has shown how deficient cavalry is, how it has fallen off instead of improving, and how much is required to be done to render it as useful to the state and as formidable to an enemy as it should be.

When I was in India an engagement between a party of the Nizam's irregular horse and a numerous body of insurgents took place, in which the horsemen, though far inferior in numbers, defeated the Rohillas with great slaughter.

My attention was drawn particularly to the fight by the doctor's report of the killed and wounded, most of whom had suffered by the sword, and in the column of remarks such entries as the following were numerous:—

"Arm cut off from the shoulder.

"Head severed.

"Both hands cut off (apparently at one blow) above the wrists, in holding up the arms to protect the head.

"Leg cut off above the knee," &c. &c.

I was astounded. Were these men giants, to lop off limbs thus wholesale? or was this result to be attributed (as I was told) to the sharp edge of the native blade and the peculiar way of drawing it?

I became anxious to see these horsemen of the Nizam, to examine their wonderful blades, and learn the knack of lopping off men's limbs.

Opportunity soon offered, for the Commander-in-Chief went to Hyderabad on a tour of inspection, on which I accompanied him. After passing the Kistna River, a squadron of these very horseman joined the camp as part of the escort.

And now fancy my astonishment!

The sword-blades they had were chiefly old dragoon blades cast from our service. The men had mounted them after their own fashion. The hilt and handle, both of metal, small in the grip, rather flat, not round like ours where the edge seldom falls true; they all had an edge like a razor from heel to point, were worn in wooden scabbards, a short single sling held them to the waist-belt, from which a strap passed through the hilt to a button in front, to keep the sword steady and prevent it flying out of the scabbard.

The swords are never drawn except in action.

Thinking the wooden scabbards might be objected to as not suitable for campaigning, I got a return from one of these regiments and found the average of broken scabbards below that of the regulars, who have steel ones. The steel is snapped by a kick or a fall; the wood being elastic, bends. They are not in the man's way; when dismounted they do not get between his legs and trip him up; they make no noise—a soldier on sentry of a dark night might move about without betraying his position to an enemy by the clanking of the rings against the scabbard.

All that rattling noise in column which announces its approach when miles off, and makes it so difficult to hear a word of command in the ranks, is thus got rid of; as well as the necessity of wrapping straw or hay round the scabbards, as now customary when engaged in any service in which an attempt is to be made to surprise an enemy.

"The cavalry steel scabbard is noisy, which is *bad*; heavy, which is *worse*; and it destroys the weapon's sharp edge, which is *worst*. The native wooden scabbard is best."—Gen. Sir C. Napier.

An old trooper of the Nizam's told me the old broad English blades were in great favour with them when mounted and kept as above described: but as we wore them, they were good for nothing in *their* hands.

I said, "How do you strike with your swords to cut off men's limbs?"

"Strike hard, sir!" said the old trooper.

"Yes, of course; but how do you teach the men to use their swords in that particular way?" (*drawing it.*)

"We never teach them any way, sir: a sharp sword will cut in any one's hand."

Had our men worn arms like these in the last Sikh war, the enemy's horsemen would not have met them with such confidence in single combat; their trenchant blades would have inspired respect—the use of them would have carried terror into the ranks of the foe.

It is well known that, beyond the effect of the moment, severe wounds inflicted in action have a depressing moral effect on the enemy.

In a pamphlet published in Berlin, on cavalry matters, it is stated that, in 1812, the wounds inflicted by the Russian horsemen inspired such *awe* that nothing but the point of honour and *esprit-de-corps* could bring the Prussian horse to close with them.

Captain Fitzgerald, of the 14th Dragoons, received a sword-wound at Rumnugger, from the effects of which he died. A Sikh, on foot, crouched under a shield, cut at him from behind. The sword exposed the spinal marrow, entering the skull at the same time.

A huge dragoon, of the regiment, was found quite dead: his head had dropped forward from a cut on the back of the neck, which had severed the spine; and at this very action, "*it is said*," that, whilst our poor fellows laboured in vain to draw blood, a touch from the Sikh's sword across the arm or leg left the bold Englishmen at their mercy, and they soon hacked them to pieces.

One officer, who was in the campaign, said he saw an English dragoon putting his hands to the reins to try and turn his horse, when a Native horseman, dropping his sword across them, took off both hands above the wrist.

A Sikh, after the retreat of our cavalry at Chillianwalla, galloped up to the horse-artillery, cut down and killed the two men on the leading horses of the gun, one after the other, and approached the third, a cool fellow, who, seeing how badly his comrades had come off with their swords, instead of drawing his, stuck to his whip, with which he flogged off his assailant's horse, and thus saved himself!

A squadron of the 3rd Dragoons, under Major Unett, charged a goel of Sikh horsemen, and the Major himself told me that they opened out, giving just sufficient room for our squadron to enter. The dragoon on the left of the front rank, going in at the charge, gave point at the Sikh next him; the sword stuck in the lower part of his body, but did not penetrate sufficiently to disable him; so the Sikh cut back, hit the dragoon across the mouth, and took his head clean off.

Colonel Steinbach, in his *History of the Punjaub*, tells us that the cavalry of the Sikh army is very inferior in every respect to the infantry; that, while the latter are carefully picked from large bodies of candidates for the service, the former are composed of men of all sorts, sizes, and ages, who get appointed solely through the interests of the different sirdars.

They are mean-looking, ill dressed, and, as already stated, wretchedly mounted. The horses are small, meagre, and ill shaped, with the aquiline nose which so peculiarly proclaims inferiority of breed.

How comes it, if our system is good, that such men, of less physical and moral courage, mounted on such inferior animals, should have been able to cope with our English dragoons? and *not seldom* successfully, for at the battle of Chillianwalla they tell of a Sikh horseman challenging the English to single combat and unhorsing three dragoons (the first, a lancer, had the lance-pole severed and his fore-finger taken off at one blow) before he was SHOT down!! And what does a charge resolve itself into, when the enemy are bold, but a *mêlée* or a series of single combats?

Let us contrast the two following gallant actions, both fought by English light dragoons; the first by three troops of the 15th Hussars against 500 French horsemen, organized and armed on the model system; the second by two troops of the 3rd Light Dragoons, against 500 of these badly-mounted rabble of the Sikhs.

In the first action the 15th charged twice.

In the second the 3rd Dragoons charged through, but the Sikhs opened out to let them back again.

Thus the 15th Hussars were twice in a mêlée with the French horsemen.

The 3rd Dragoons only once with the Sikhs.

The First.— "In the general attack made on the 2nd of October (1799) on the positions occupied by the enemy at Bergen and Egmont-op-Zee, the 15th Dragoons formed part of the cavalry under Colonel Lord Paget (now Marquis of Anglesey), attached to the force commanded by Sir Ralph Abercromby.

"Moving along the sea-shore towards Egmont-op-Zee, this column encountered a strong force of infantry among the sand-hills, with a numerous body of cavalry and artillery, to their left on the beach; when a severe contest ensued, in which the steady bravery of the British infantry triumphed.

"After forcing the enemy to fall back for several miles, the cavalry advancing along the beach, as the infantry gained ground among the sand-hills, the column halted in front of Egmont. The British artillery moved forward to check the fire of the enemy's guns, and two troops of the 15th Light Dragoons advanced to support the artillery. Lord Paget posted the two troops in ambush behind the sand-hills; and the French General, thinking the British guns were unprotected, ordered 500 horsemen forward to capture them. The guns sent a storm of balls against the advancing cavalry; a few men and horses fell, but the remainder pressed forward and surrounded the artillery. At this moment the two troops of the 15th sallied from their concealment, and, dashing among the assailants, drove them back upon their reserves, and then returned to the liberated guns.

"The opposing squadrons having rallied, and ashamed of a flight before so small a force, returned to the attack. They had arrived within forty yards of the 15th, when a third troop of the regiment came up, and a determined charge of the three troops drove the enemy back again with loss, the 15th pursuing above half a mile.

"The loss of the 15th was *three men and four horses killed*; Lieut-Colonel James Erskine, nine men, and three horses wounded.*

The Second.——The cannonade had not been of long duration, when a body of Sikh horsemen moved to Sir Joseph Thackwell's left flank, as if to get into his rear. He now ordered the 5th Native Light Cavalry (three squadrons), and the Gray squadron of the 3rd Dragoons (Unett's), to charge and disperse them.

The gallant General had a lively recollection of what a few squadrons of British dragoons effected in the Peninsula against the French, and reasonably entertained the expectation that the force would prove suffi-

*Historical Record of the 16th Hussars.

cient to drive back the Sikh irregulars. It was his intention to have advanced the few men left at his disposal, namely, the remaining squadron of the 3rd Dragoons and 8th Light Cavalry (native), on Outar's force, as soon as the success of the other charge became manifest. The charge was sounded, and Unett's squadron, in line with the 5th Native Light Cavalry, approached the enemy. The Sikhs commenced a desultory matchlock fire.

Unett steadily advanced, hut the 5th, put into confusion by this reception, went about and fled in the greatest precipitation, in spite of the most earnest entreaties of their officers, of whom several received wounds.

The 3rd, forcing their way through the hostile ranks, never pulled rein till they had got some distance beyond the enemy. Unett, who was severely wounded, found his men sadly dispersed.

The few men around him, with clenched teeth, essayed to cut their way back. The Sikhs opened out, and, giving the dragoons a passage through them, abused, spat, and cut at them.

The other parties under their officers, the gallant Stisted and Macqueen, repassed the enemy as they could.

The *casualties* in this squadron were not less than *forty-six*.

The suspense of every one was great; Sir Joseph himself became apprehensive that the squadron was annihilated.*

Here we see two troops of English dragoons dashing into the middle of 500 victorious French horsemen, and, after a mêlée, driving them off.

We then see these same 500 French horsemen returning boldly and meeting the English (now reinforced by one troop) at the charge; a second mêlée ensues, and in both conflicts they only kill three Englishmen; and these French dragoons are men well mounted, of undoubted courage, disciplined and trained according to our own system.

In the *second* instance, two troops of the 3rd Dragoons charge with great gallantry and break through the enemy's ranks; when charging back the Sikhs open out and let them through; in the mêlée with these men, so mean-looking and wretchedly mounted, they lose 46 men and horses, and nearly lost their gallant leader, Major Unett, who received

The Second Sikh War, by E. J. Thackwell. Battle of Chillianwalla

a sword-cut which divided his pouch and entered two inches deep into his back.

Another comparison of this kind may be of use, if only as an additional impress to the memory.

At the battle of Heilsberg, on the 18th of June, 1807, a good fight took place between a division of the French Cuirassiers d'Espagne and two regiments of Prussian horse; one a regiment of lancers, the other the dragoons of Ziethen. The French cuirassiers met the Prussians at a walk and at close order: a hand-to-hand fight ensued, which terminated in favour of the Prussians, who drove back their opponents into the wood at Lavden. In that well-known French work *Victoires et Conquêtes* it is mentioned that a French officer came out of this fight with *fifty-two* new wounds upon him, and that a German officer, Captain Gebhardt, received upwards of twenty wounds. It appears that Captain Gebhardt did *great* execution with the shaft of a broken lance, knocking several cuirassiers off their horses, but that he was himself put hors-de-combat by a kicking horse which rolled him over! Imagine a man receiving *fifty-two* sword and lance wounds without losing life or limb! No wonder the Prussian Gebhardt took a big stick at last (and a broken lance-shaft is only a big stick) as the most formidable weapon within reach—and by far a better tool than such sabres.

I have little to say about helmets, caps, jackets, and dress in general, except that the most simple are the best. This opinion is now gaining ground universally. But by simplicity I do not mean roughness, shabbiness, or inelegance. So long as you can keep a soldier, let him be well dressed and smart in his appearance. The greatest element of true elegance is the very simplicity which I would recommend. To me it appears that we have too much frippery—too much toggery—too much *weight* in things worse than useless. To a cavalry soldier every ounce is of consequence. I can never believe that our hussar uniform (take which of them you please) is the proper dress in which to do hussar's duty in war—to scramble through thickets, to clear woods, to open the way through forests, to ford or swim rivers, to bivouac, to be nearly always on outpost work, to "rough it" in every possible manner. Of what use are plumes, bandoliers, sabretashes, sheepskins, shabraques, &c.?

"It seems decreed that the hussar and the lancer is ever to be a poppinjay—a show of foreign fooleries, so laced, and looped, and braided, that the uninitiated bystander wonders how he can either get into his uniform or come out of it. A woman's muff upon his head, with something like a red jelly-bag at the top, has been substituted for the warrior's helm; and the plume, so unlike the waving horse-hair of the Roman casque, would seem better fitted for the trappings of the undertaker than the horseman's brow. The first time I ever saw a hussar, or hulan, was at Ghent, in Flanders, then an Austrian town; and when I beheld a richly decorated pelisse, waving empty, sleeves and all, from his shoulder, I never doubted that the poor man must have been recently shot through the arm; a glance, however, upon a tightly braided sleeve underneath, made it still more unaccountable; and why he should not have had an additional pair of richly ornamented breeches dangling at his waist, as well as his jacket from his shoulder, has, I confess, puzzled me from that time to the present, it being the first rule of health to keep the upper portion of the body as cool; and the lower as warm, as possible. Surely a horseman's water-proof cloak, made to cover from head to foot the rider and his saddle, with his arms and ammunition—to be his protection against the pouring deluge, his screen and cover in the night bivouac—is the only equipment of the kind the country should be called upon either to furnish or suffer.

"Man-millinery in any shape is an abuse and prostitution of the English character. Borrow and copy from foreigners whatever may be truly valuable in arms—it is right and fitting so to do; but let us dress ourselves in serviceable garb, that *fears no stain, nor needs a host of furbishers to keep in order.*"

Of the tight leather stock and head-piece Dr. Fergusson says—

"The circulation of the ascending arteries in the neck is by far the closest of any part of the human body, and to impede its relief by the returning veins, which a stiff ligature of any kind is sure to do, must have a stupifying effect upon the brain.

"It cannot fail, besides, to deteriorate the sight, from the pressure of congested blood upon the optic nerve, and the stock would seem to be preserved only for the purpose of generating a tendency to all kinds of apoplectic and ophthalmic diseases.

"It would be better, surely, to inflict an ulcer upon the soldier's neck, for the discharge might then have the relieving effect of an issue; but a tight ligature, not only on the neck, but anywhere else, should be rejected for ever from military dress and equipment of whatever description.

"A heavy head-piece is everywhere a disqualification and a hindrance to the wearer, for, to heat and cumber the brain, which, being the source of all our powers and faculties, ought ever to be freest, can never be justified.

"Everywhere the direct rays of the sun striking upon the eye must be hurtful; but when these are refracted from a white rocky soil, the immediate effect becomes distressing in the greatest degree. A shade properly dropped from the cap would effectually obviate this, and it ought to be furnished."*

Without defensive armour (which brings weight to the horse and cramps the man), a good uniform may afford valuable protection, while a bad one may be as inconvenient as armour, and afford no protection where it is most wanted.

---

*These extracts are from a book called *Notes and Recollections of a Professional Life*, by the late William Fergusson, Esq., M.D., Inspector-General of Military Hospitals, a work so instructive and interesting to all military men that I recommend it strongly to their notice. It is edited by his son, the well-known James Fergusson, a man of distinguished ability in many ways, and one who, of late, has stood prominently before the public, and created no small sensation by his last work, *The Perils of Portsmouth*, and by his new system of Fortification.

In speaking of the foreign hussar, Dr. Fergusson was apparently not aware that the Hungarian hussar is actually provided "with an additional pair of richly ornamented breeches," which are worn under the overalls on the line of march; at least they were so provided when I had the honour of serving with them. Round the neck they wore a loose black handkerchief, which I should earnestly recommend for imitation: in the copy we have made of their dress a stock has been substituted. The Hungarians line their pelisses with fur, and make them like a peajacket to go over the dollmann (dress-jacket) in cold weather; our pelisses are not large or loose enough for this purpose; we therefore generally sling them, or wear them instead of the jacket.

A squadron of the 15th Hussars in India marched into the Nizam's country in 1850, where the slung pelisses created the utmost curiosity amongst the natives at Hyderabad, and gave rise to many stories as to their origin. The prevailing belief, however, was, that the regiment had done such gallant deeds on one particular occasion, that the king said "each individual fought as if he had four arms," and gave them four arms accordingly—a comfortable belief, which we took care to leave them in.

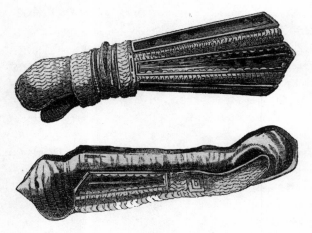

FIGURE 1. GAUNTLETS.

The most vulnerable parts of a cavalry soldier are the head, the back of the neck, and the arms and legs. The Asiatics, well aware of this, cover those parts in their own bodies, and take immediate advantage of those who are not protected in like manner. The Turkish irregulars wore a turban, that was a better defence to the head than our helmets of brass or of steel: with them the leg was defended by the deep saddle and the broad shovel-like stirrup-iron; the arms of their jackets or beneeshes were padded, and the blunt European sword seldom cut through the dress, which was of silk, or of silk interwoven with cotton. The jacket of the Russian or Austrian trooper offered no such impediments to their scimitars, and short, handy, light yataghans.

"The propensity of the Sikhs to aim their cuts at the back of the head was so unequivocally manifested on the 22nd of November, that it became an object of consideration to the officers of the army to provide some defence, however slight, for the precious *caput*.

"Some officers wrapped rolls of linen cloth round the back of the csako, the folds of which hung down over their backs, affording some protection."*

*Thackwell, *History of the Second Sikh War*.

For the arms, gauntlets ought to be used, which would leave the hand free and naked to grasp the sword; like those in use with the natives of India, made of steel, to reach to the point of the elbow (*Figure 1*). Arms thus encased can be used to parry blows aimed at the head or body—the gauntlets are no weight, and the hands being free and naked gives a great advantage in action.

Overalls afford no protection in the ranks or in the mêlée; they are constantly getting torn, the lower parts get rotten with mud and wet, the straps impede the bending of the knee in mounting, and prevent a man from exerting himself when on foot.

Overalls without straps, and black leather leggings, made the shape of Napoleon's boots, with a covered bar of steel running down the side, fastened with straps above and below, and only worn when mounted, would give the horseman the necessary protection and enable him to exert himself as long as life was in him; for a chance blow over arm or leg would not disable him.*

Colonel Ponsonby was thus disabled at Waterloo, and gives the following account of himself:—

"In the mêlée I was almost instantly disabled in both my arms, losing first my sword and then my rein; and, followed by a few of my men, who were presently cut down, no quarter being asked or given, I was carried along by my horse, till, receiving a blow from a sabre, I fell senseless on my face to the ground."†

With regard to fire-arms for cavalry, as accuracy of fire is what is required, not rapidity, I should give them short handy rifles: these should be carried in a holster, about 14 inches long, bell-mouthed, like a tube, and open at the bottom. This holster should be fastened below the off wallet, pointing to the horse's shoulder. The carbine is pushed through, and a strap about one yard long fastens it to the pummel of the saddle, and prevents the man from losing the weapon if he should let go his hold when firing or loading.

---

*When dismounted the dragoon could leave leggings and spurs with his horse, and turnout in ankle-boots and trousers without straps.

†Cotton, *A Voice from Waterloo*.

The cumbrous bandolier-belt with its appurtenances is got rid of,* and the man can bring his carbine easily to his shoulder; whereas it is almost an impossibility to do either whilst the carbine is attached to the big awkward belt. No carbine stay-strap is required and no bucket: the carbine can be drawn and returned with the same case as a pistol. It rides in much the same position as at present, with the bucket and stay-strap, only it rides steadily and does not get into the way of the man's sword-arm, which happens often at present when jerked forward by the stay-strap. The strap I propose unbuckles from the saddle and serves as a sling when the man is dismounted.

Another good way of carrying the carbine is to sling it at the back, muzzle over the left shoulder, the small of the butt fastened to the waist-belt on the right side by a small strap and button to keep the carbine steady. Thus the soldier has his weapons about his person, and if unhorsed can effectually defend himself; and the carbine would often, in action, save the wearer from a sword-cut across the back.

Many foreign regiments carry the carbine hooked to the bandolier, muzzle downwards, and let it dangle about loose. It is terribly in the way of the men's legs in the ranks; when galloping, the butt strikes against the hip-bone, occasioning great suffering; encumbered with the belt and hook, the man can never bring the weapon freely to his shoulder to take aim, and it is most inconvenient in mounting and dismounting.

It appears to me that the distances for which they are now sighting soldiers' carbines abroad are quite ridiculous. Few have eyes good enough to see a man or even a column so far off. This custom, if not altered, will lead to the men firing at everything they fancy they see in the distance, causing constant false alarms in camp. No supplies of ammunition will suffice for troops firing in this manner. It may be very well for some foreign troops to stand out at such long shots, and so keep danger far off; but I hope and trust that an honest Englishman will always like to look his enemy nearer in the face. I think that cavalry carbines sighted up to 300 yards would more than suffice for all purposes.

---

*The pouch-belt, from the manner in which it is slung across the man's back, shakes the ammunition to pieces, and soon renders it totally unfit for use. The ammunition should be in a waterproof box, fastened to the off-side of the saddle, in lieu of the present wallet, the soldier carrying six rounds in his waist-belt for immediate use.

One of the greatest difficulties of the officer is to make his men reserve their fire. Surely nothing should be done to increase this difficulty, or to tempt the men to such long shots.

The Lance and the Sword

Formerly it was a received opinion that the lance was particularly formidable in single encounters, that the lancer should be a light, active horseman, and that space was required whereon he might manage his horse and turn him always *towards* the object at which he was to thrust. But of late there seems to be rather a disposition to take up Marshal Marmont's notion of arming heavy cavalry with lances, to break infantry as well as cavalry. All seem to forget that a lance is useless in a mêlée,— at the moment the lancer pulls up and the impulsive power is stopped, that instant the power of the weapon is gone.

The 16th Lancers broke into the Sikh squares at Aliwal, and in the mêlée that ensued these brave men attacked the lancers *sword* in hand and brought many of them low, for they could effect nothing with the lance.

In the second Sikh war, I have been told that our lancers often failed in driving their lances into a Sikh, because they had shawls wrapped round them. I could tell them a better reason: it was because those who failed did not know that it requires speed to drive a lance home, and that it must be carried into the object by the horse.

I have often seen, when hog-hunting, men with spears sharp as razors unable to drive the weapon through the boar's hide, whereas others (old hands) would send a spear in at one side and out at the other, through bone and all.

This shows that the lance is not a dangerous weapon in all hands, and therefore unfit for soldiers.

All experiments with blunt lances on fresh horses go for nothing, in my opinion, for many of the thrusts would not go through a man's jacket; and in a campaign, when horses are fatigued, and will not answer the spur, even the skilful horseman is helpless with a lance in his hand.

At speed you can drive a lance through anything, but not so at a slower pace; and at a walk, and a stand, you become helpless, and the thrust can be put aside with ease, or the pole seized with the hand.

If the advantage of the lance is in its long reach, the longer the weapon the more formidable. The French gensdarmes, whose lances were 18 feet long, suffered such dreadful defeats that they gave up the weapon altogether.

Gustavus Adolphus took the lances away from his cavalry in the Thirty Years' War. He had practically experienced their inefficiency.

Let us allow, for the sake of argument, that a lance of a proper length, handy, well poised, and held at its centre, reaches further beyond the horse's head than the point of a sword held at arm's length: in what way can this conduce to success, when it is universally acknowledged that it is the superior impetus and speed of one of the advancing lines which overthrows the other; the weapons only coming into play afterwards?

The lancers' pennons attract the fire of artillery; in single combat they betray to the adversary where the danger is, and thus enable him to avoid it; and if they sometimes frighten an adversary's horse, the animal shies and carries his master out of reach of the point which, if not decorated, might have run him through the body.

The Asiatics carry a light spear (without pennon), which they say they leave in the body of their first foe (or throw away), but take to their swords when the tug of war comes fast and fierce.

I believe that the only advantage of the lance lies in the moral effect produced (particularly on young soldiers), not only by its longer reach, but by the deadly effect of the home thrusts. Thus in the Seven Years' War, the Prussian hussars were at first very shy of the lances used by the Russians: some of the Prussian officers rode out in front of the line engaged, and cut down several of the Cossacks and lancers in single combat, showing their men how easy it was to despatch them by closing upon them at once; and thus encouraged, the hussars soon mastered their opponents.

The Russian line of outposts, formed against the Circassians is inhabited by the Line Cossacks, though many Cossacks of the Don and Ural are there on service. All form part of the army of the Caucasus.

Those of the Don, the Ural, and Tschernomor, are armed with the lance; but the Line Cossacks, who are in constant warfare with the Circassians, have given it up, and taken to the sword. They say that in irregular warfare the lance is useful against bad horsemen, but that it is only in the way in a contest with bold and skilful riders like the Circassians, who close upon them at once.

These Cossacks contend, and often successfully, against their adversaries; but the Circassians, with their swords, make short work of the other Cossacks, and look upon the Line Cossacks alone as formidable antagonists. As to the lance being a more deadly weapon than the sword, it depends entirely upon what swords it be compared to. If to those generally in use in the European cavalry of the present day, decidedly we must give it in favour of the lance; but if we compare it to the scimitar of the Turk, the sword of the Mahratta, the Sikh, the Circassian, any sword with an edge to it, then, I say, the lance is comparatively but a harmless weapon.

If lances be such good weapons, surely those who wield them ought to acquire great confidence in them, whereas it is well known that, in battle, lancers generally throw them away, and take to their swords. I never spoke with an English lancer who had been engaged in the late Sikh wars that did not declare the lance to be a useless tool, and a great incumbrance in close conflict.

We often attribute want of success or defeat to the tactics of an enemy, or to his arms being superior—anything rather than acknowledge that his more manly courage won the day.

Thus, perhaps, have the Line Cossacks thrown away their lances and taken to the sword, to imitate their more successful enemies, the Circassians. And the Prussians in 1813 copied the Russians, and adopted the lance for the landwehr.

In the last Hungarian war the Hungarian hussars were (as we have seen) generally successful against the Austrian heavy cavalry— cuirassiers and dragoons; but when they met, the Polish lancers, the finest regiments of light-horse in the Austrian service, distinguished for their discipline, good riding, and above all for their *esprit-de-corps* and gallantry in action, against these regiments the Hungarians were not successful, and at once attributed this to the lances of their opponents. The Austrians then extolled the lance above the sword, and armed all their light cavalry regiments with it.

Russia has armed the front ranks of her heavy cavalry with long unwieldy lances, and other European powers have lately been following the example. Marshal Marmont, an exile from France, was in Russia when he took up his new idea.

The failure of the 7th Hussars in the retreat from Quatre-bras, against the French lancers, jammed close together in the streets of Gemappe, was attributed to the lances of their opponents.

Of what use were the lances to the French a few minutes later, when a regiment of life-guards (without cuirasses) went at them sword in hand, and drove them through the town, and out at the other side—riding them down, and cutting them from their horses in all directions.

Yet after the battle of Waterloo, lancer regiments, for the first time, were formed in England!

Lancers are of no use for outpost duty; the enemy shoot them down, and they have no fire-arms wherewith to keep the enemy off.

The French lancers in attacking lancers of other nations often slung their lances, and drew their swords. General de Brack recommends swordsmen, engaged with lancers, to close upon them, and crowd them together. He says, "The lancers, jammed together, can neither point nor parry, and one of two things must happen; they will either throw down their lances, in order to get at their swords, or they will retain their lances, and in this case you will have the best of the bargain. Our pivot files in the lancers of the Imperial guard did not carry lances. I remember upon two occasions in 1814 (at Hoagstraten, near Breda, and at Pont-à-Trecir, below Lisle) having to deal with Russian and Prussian lancers, who, like ourselves, held their own upon narrow roads with deep ditches on either hand. I placed carabineers at the head of my column, and made my lancers follow; and these last put their lances in the bucket, and drew their swords; and, having penetrated the enemy, our success so far surpassed our hopes, that we cut them down without damage to ourselves."

A good plan, however, is often followed by the Cossacks when attacked by swordsmen. They stand last and receive the assailant on their left, keeping the lance to the right front: when the swordsman is within reach, they make a circular parry towards him (from right to left), and turn the assailant aside by this movement. They then turn to the left and follow, and charge home with the lance at their opponent's left side.

The Poles, by constant habit and practice, may possess a peculiar aptitude in the management of their national weapon, but with the rest of the European nations the lance may be said to have had its reign. At least few will now say, with Montecuculi:—*"La lance est la reine des armes blanches!"*

The following extract from a letter of Oliver Cromwell, giving an account of the battle of Dunbar, shows clearly that arming the front ranks of cavalry with lances is but an old custom revived:—

"The dispute on this right wing was hot and stiff for three quarters of an hour. Plenty of fire, from field-pieces, snaphances, match-locks, entertains the Scotch main-battle across the Brock; poor stiffened men, roused from the corn-shocks with their matches all out!

"But here on the right, their horse, *with lancers in the front rank*, charge desperately; drive us back across the hollow of the rivulet;— back a little; but the Lord gives us courage, and we storm home again, horse and foot upon them, with a shock like tornado tempests; break them, beat them, drive them all adrift."*

Where shall we find, in these our modem days, a bulletin shorter, and better, than this?

The Horse

I assume that in England, in Ireland, in Canada, in India, at the Cape of Good Hope, and in Australia, we have always the means of obtaining good remount horses for our cavalry. But as one may make something of a horse of indifferent quality by skill, so the very best horse may be spoiled, or greatly injured, by a bad system of accoutring and training him. For the present I speak only of bridling and saddling.

Here nothing is so important as the selection of the proper bit. This is a subject upon which many have been very positive and pedantic, and few very wise.

The art of suiting to each horse a bit of more or less power, according to the shape of the mouth, the sensitiveness and the temper of the

*Carlyle's *Cromwell*.

animal, is looked upon abroad as a science; and in Prussia it is said to stand higher, and to be more difficult to learn, than riding itself.

They say that the best broken horse would be ruined if a bit was put in his mouth which did not suit; and so far is this carried by Germans, that they will tell you seriously that to shorten the curb-chain *one* link, or to use a bit half an inch longer or shorter in the cheek, will make the difference between a well-broken horse and a restive one.

What, then, does a good hand go for?

The Arab and the Turk ride with bits so powerful that they can break a horse's jaw or pull up.

The Persians, Sikhs, and Mussulmans, have a square snaffle, with iron spikes; they wrap thread round this mouth-piece, so that, the more the horse presses against it, the farther the spikes come out from under their covering, and run into him; and farther, they have a standing martingale, fastened to the rings of the snaffle, which keeps the horse's head down and under control.

The Cossack and the Circassian, the latter particularly famed for the wonderful address with which he handles his horse at speed, and in single combat, both use a common snaffle.

It is not the shape of the bit, the horse's mouth or temper, nor is it the nation and peculiarities of the horseman, that render the animal obedient and handy, but it is the way of riding and breaking him, and the manner in which you teach the horse to obey the bit you put in his mouth, whatever that may be.

A bit of moderate power is the best for general purposes, and if properly used will bring most horses under control.

The mouth-piece should be sufficiently arched to admit of the horse's tongue passing freely underneath it, and the cheek rather long, to give power to the lever.

But, beyond the consideration of a mouth-piece, there is another quality of essential importance in the bridling of cavalry. It is to afford the possibility of bridling and unbridling quickly; for if there is difficulty in bridling; up, the soldier cannot feed the horse in the near neighbourhood of the enemy, and this, in a campaign, becomes matter of serious import, for the condition of the horses will suffer. The Cossacks feed their horses at all times, even in the battle-field, amidst the roar of cannon. Their horses kept condition in the last war, when others were dying from exhaustion and want of food, and the marches they sometimes made were wonderful.

The Russians have their bits made with a hook and link, and can take them out of their horse's mouth without taking off the bridle. (*Vide* illustration in the Appendix.)

Some of the bridles in use in our cavalry (those of the Carabineers and the Inniskillens) are, in my opinion, better than those of the Russians. They are on a more simple plan, and, with the Russian link, would afford the same facility of bridling quickly.

The snaffle, however, should have half-horns, to prevent the rings from being drawn into the horse's mouth; and the reins should be sewn on, for the strain is then on the breadth of the rein, instead of the tongue of the buckle.

Ought not this convenient bridle to be at once made general in the service? In a campaign it obviates the serious evil which I have indicated. But, without going to the field of war, everybody in these days of railway speed feels the necessity of economising time and trouble. One has often to bridle in a hurry, and sometimes, it may be, in the dark.

The Saddle

That saddle is best for cavalry which, being of a simple construction, brings the soldier close to his horse, in a firm and easy seat. It must be sufficiently strong to carry the necessary kit, easy to repair or replace in the field; it ought to be roomy, and, above all, it ought to give no sore backs.

I have cited the "Magnanimous Usurper" as an authority, and such he must be considered in cavalry matters. The following brief, pointed, and truly characteristic letter, will prove the attention he bestowed on horse-gear and sharp swords:—

"Wisbeach, this day, 11th Nov. 1642.
"Dear Friend,

"Let the saddler see to the horse-gear. I learn from one, many are ill served. If a man has not good weapons, horse, and

harness, he is as nought.
"From your Friend,
"Oliver Cromwell
"To Auditor Squire."

Every cavalry officer will do well to remember this curt epistle.

The present hussar saddle raises the man high off his horse, because the spread of the side-boards, and the upright position of the forks, require the wolf, or seat, to be high; for if it is not high, then the sides of the boards raise up the thighs of the man, and prevent him from gripping his horse. The saddle is like a wedge between the man's legs, on which his body acts as a lever: thus, if he inclines or throws his weight on one side or the other, he moves the saddle bodily, and by doing this often he naturally causes a sore back.

In this respect a great improvement has been made in the Hungarian saddles now given to our heavy dragoons; for here the side-boards are cut out like the tree of our English saddle, the seat is lowered, and the. man brought nearer his horse.

The Hungarian saddle is used either with a large blanket in twelve folds, or with pads stuffed with horse-hair. Now both these methods are faulty. The large blanket is, in the first place, very hot and oppressive, the side-boards, being of a polished surface, have no hold on the blanket, and thus it often works its way out from under the saddle. With pads, a free circulation of air is indeed obtained between the saddle and the horse's back; but as the condition of the horse varies, so does the position of the saddle; and as this cannot easily be rectified by means of the blanket, different expedients are resorted to, such as covering the side-boards with plaits of straw, or giving the blanket an extra fold.

Every one must have had occasion to observe, on a long march, that the edges of the pads stretch the horses' skins, and occasion sore backs; then they become hardened with the sweat from the horses, and gall them severely.

Instead of pads, I would recommend the use of strips of felt, put in a cover, and laced on to the side-boards. For the large blanket I would substitute a felt saddle-cloth, to reduce the direct strain on the horse's skin, absorb the perspiration, and prevent the edges of the pads from getting hard. The slips of felt enable you always to keep the saddle in its proper position; for, if a horse falls off in flesh, you place an extra slip of felt in the pad, and, if he puts up condition, you take one out.

I have seen a trial made with a squadron that had pads and many sore backs. They then used a blanket under the pads, nearly in the manner I have recommended, and in spite of forced marches the sore hacks got well, and the squadron, after travelling 400 miles, arrived at head-quarters with only one led horse.

The Turks very generally use felt, in the manner here described. Two recent English travellers found it impossible to get on in Asia Minor with their English saddles, without bringing in the aid of strips of felt; but with this aid they made several long journeys without once galling the backs of the very poor creatures they rode. Good felt was cheap in the country. They carried a supply of strips with them, and these they applied in increase, or removed, or changed, as the occasion or the condition of the horses required. With a plentiful supply of this material, a Turkish or Greek peasant, with a clumsy pack-saddle made entirely of wood, will ride a horse without soring his back, or wringing, or galling him.

In Hungary, the peasants use their wooden saddles without any blanket. In the French service they have been making experiments to dispense with the blanket altogether. This, no doubt, might answer very well in time of peace, or so long as the horses can keep up their condition; but the moment the wood comes in contact with the ribs the horses must suffer, and then a very short time will suffice to wear through their skin.

A very ingenious saddle has been constructed and tried in Belgium: it has moveable side-boards, the front and hind forks working on an iron roller, so as to allow of the side-boards always assuming a position parallel to the horse's back. But, unfortunately, this saddle is very liable to get out of order, and the kit does not pack so firmly on it as it ought to do.

For details about the saddle of my own invention, I must refer to the Appendix of the present volume.

The seat for a horse-soldier should certainly lie low, like that of a hunting saddle, and the saddle should be broad and roomy, the front part flush, the side-boards closer in front. If I am not deceived, the saddle of my proposed construction will meet all the desiderata, and will be a comfort and a security both to man and horse.

6

MILITARY RIDING

ORSEMANSHIP, LIKE THE NOBLE animal itself, found its way into
Europe from the East. Though no longer for war, nor even for the
manly sports of the field, the art was highly cultivated during the
decline of the Eastern Empire. The hippodrome of Constantinople
attracted the attention of Europe when no other object of admiration or
of sympathy was left in that demoralizing and falling state. About the
year 1134 many companies of Byzantine circus-riders went over to
Naples, at that time the last state in Italy which acknowledged the
authority of the Eastern Empire. Thus Naples became the first school
for horsemanship in Western Europe. From that city the school was
gradually spread over the rest of Italy, and into France and Germany. For
some ages Naples also supplied the best horses for the manège. Down
to the close of the sixteenth century, and even to a later period, the
Neapolitan horses are frequently mentioned by writers as highly prized
in England as well as in most of the continental countries. They divided
favour and pre-eminence with the well-bred horses imported from the
south of Spain, where the blood of the Arab and the Barb had been lib-
erally infused. As stock, they tended to improve the studs of other countries. It should seem that, by importation and by other means, England
must have had a certain supply of good nags in the fourteenth century,
or as early as the time of Chaucer, for that old poet frequently alludes
with evident gusto to choice horses and neat horsemanship. Of his lord-
ly sporting Monk he informs us that—

"Full many a dainty horse had in stable."

Of his aged Knight he tells us that—

"His horse was good, albeit he was not gay."

Among the accomplishments of his young Squire he does not forget to admit that—

"Well could he sit his horse and fairly ride."

But the most charming figure that rode with that good and merry company of pilgrims to the shrine at Canterbury was the Wife of Bath—

"Upon an ambler easily she sate," &c.

Our early school of horsemanship was certainly easy and natural, and therefore good. As such it may be said to have continued—except for the cavalry of our army. Whatever ease originally belonged to the Neapolitan school (which, being of Eastern origin, was doubtlessly good) was soon stiffened out of it in France and Germany, as also in every part of Italy.

The Ironsides of Cromwell kept their national and natural seats, and rode on the field of battle as they would have ridden across country. But France and Germany, by the time of Marlborough, came to be considered as the great emporia of military science, and thither repaired all young Englishmen who aspired to glory and renown in arms, to study the practice as well as the theory of war. Now the French have never been an equestrian people, and the Germans must certainly yield the palm of horsemanship to the English. Moreover, both French and Germans fell into a very stiff and artificial mode, introducing pedantry into the riding-school, and depriving both man and horse of their capabilities and natural élan.

Yet our *military* riding is to this day imported from the Continent.

I say *military*, because none of our dragoon or hussar officers would for a moment think of riding across country in a foreign seat, or in any other way in the manège fashion. Yet in the business of war our cavalry ought to be able to perform whatever is done in the sport of hunting, and whatever interferes with the ability of so doing must be set down as a detriment and great evil.

The Russians and most other continental nations place their saddles near the loins, and girth their horses round the belly instead of the brisket.* The established seat is upright, the knee drawn back, and the

---

*This is the order in our cavalry regulations, though fortunately it is not strictly carried out. The regulations say, "Place the saddle one hand's breadth behind the play of the shoulder."

heel in a perpendicular line with the point of the shoulder; and so far is this carried, that it is no uncommon thing to see spur-marks on the horse's stifle. The man's legs, from the knee down, are carefully brought away from the horse, in order to prevent what is called "clinging," and he is taught to ride by *balance*. If this system can be right, I have thrown away many years in the study and practice of horsemanship, and all who are bold riders at home, and all the best cavaliers of the East, are wrong.

And how do foreigners treat the horse? The system followed in their riding-schools wearies out all patience, both in man and beast. The *simple* breaking in of the horse has no end; and if his education is to be perfected by the addition of a few "airs de manège," such as croupades, pirouettes, ballottades, pésades, &c. &c., then the horse must be long-lived to be brought to execute them! And should the veteran of the manège take all his degrees with the highest credit, for what mortal work is he fit? His life has been consumed in doing unnecessary things, and he has no strength and no legs left to do things necessary and essential.

In almost all cavalry services detachments from each cavalry regiment are sent to the various riding establishments to learn to ride. These detachments are composed of picked men and horses, carefully trained before they are sent to these establishments for instruction. The detachments remain about a year, in most foreign countries still longer, and do nothing but riding-school work, for two and three hours a day, during all that time, and all this to go through a ride at a walk, trot, and canter when they are dismissed.*

Of what possible use can a system be which requires trained men and horses to ride daily in the riding-school for a whole year, to enable them to go through a ride in the school, which constitutes, after all, but the first step in their training on horseback? for the same men have never ridden once at speed, or used their arms at that pace!

The system of brute force pursued by the horsemen of the East is far preferable to this; in eight days they make their chargers canter round a sixpence, and pull them up and turn them at speed. This they do by tying the horse's nose down with a standing martingale, attached to a

*At our riding establishments they have much improved of late. They make the men perform the sword and lance exercise at a gallop, and teach them to take a flight of fences put up in the barrack-field.

spiked snaffle; they then fasten a rope to the rings of the snaffle and longe the animal on a very small circle, with a man on his back, whose spurs and whip cannot be denied. After a few days at this, they practise the horse at starting off at speed and pulling up on the spot, and their charger is ready for the field! True he can neither walk nor trot; that is, he walks generally with both legs of a side at the same time,* and instead of trotting he amities; but nevertheless great results have been obtained, for the horse is *handy* and *obedient*.

At ———, on the Continent, Z. Z. showed us the royal stables, and the horses broken in at the riding-school. One of them had no shoes on; we asked the reason. Answer. "He never works out of the riding-school." —Question. "How old is he?" A. "Fourteen years old." —Q. "Is he quite perfect in the riding-school work?" A. "Not quite, but very good at it."

We were shown a "Springer." A groom led in a horse with his tail tied on one side (I presume to give a better opening for the whip of the riding-master), a cavesson on, and a young man in jackboots riding him, his legs drawn down and unnaturally far hack, a cutting whip held upright in one hand, and the reins divided in both hands. The horse was placed against the side wall, the groom in front with the cavesson line held up to prevent the horse springing forward. The animal was evidently uneasy, and looked back. No wonder! for presently the riding-master stepped up behind, and crack! crack! went the whip into the "springer's" unprotected hind-quarters. He sprang in the air and back to his place, for he could not get forward. This was not enough. It appears that the perfection of this performance consists in getting the horse to kick out behind at the moment he is off the ground with all-fours; and, what between the groom pulling the iron band against the horse's nose with all his might, and the riding-master giving him the whip with a practised hand, they succeeded in getting the capriole required, sending the man in boots on to the horses neck at the same time.† The riding-master,

---

*This manner of walking, as well as bridle-lameness, is common amongst our school horses.

†There are two airs de manége of this sort in the old school, both equally useless. The first is called the ballottade, in which the horse jumps off the ground, bending both knees and houghs, and showing his hind shoes, without however kicking out. The second, called the capriole, is the same, only the horse lashes out whilst off the ground.

pleased with the success of his experiment, turned to us to explain how difficult it was to get a horse to do it. I asked how long had the horse been at it. "Oh," said he, "he has been a springer for several years." In fact, he was a lucky beast, and had got his *promotion* early in life.

We then saw the cuirassiers of the Guard at exercise. The men and horses were heavy and unwieldy, worked slowly and loosely, and did not come up to our expectations. What amused us greatly was that they all had their saddles so far back as to sit on the animal's loins, and looked very much like men riding their donkeys to market.

There was an enterprising medico attacked to the regiment, who, without the least regard to his own safety, took his horse several times over a *dangerous* ditch, about *two feet wide*, and when he observed us looking at him he repeated this feat over and over again, calling out "Hop!" with all his might. He little thought that we were marking the *rise and fall* on the church steeple in the distance, for the city itself we plainly saw at every jump between his seat and the saddle.

The result of this long and monotonous course of study is, that on the *uninitiated* the school-rider makes a pleasing and striking impression; his horse turns, prances, and caracoles without any visible aid, or without any motion in the horseman's upright and imposing attitude. But I have lived and served with them. I have myself been a riding-master for years, and I happen to know from experience what disadvantages are of this foreign seat and system. These I shall endeavour to explain.

The balance-seat originated in necessity. It was indispensable when combatants, sheathed in armour, ran a course with lance in rest. The upright seat enabled them to carry the weight of the armour with more ease, and the long stirrup supported the leg at that point to which the weight of the armour pressed it down. They were obliged to study balance on horseback, for, the equilibrium once lost, no effort of strength could save them; the weight of the armour brought them to the ground.

As a pole is balanced on its end by bringing the hand from side to side, backward or forward, so were these knights balanced by their horses, through the use of hand and leg. The necessity which introduced the system has long ceased to exist; but the system is kept up notwithstanding; and the riders, accustomed easily to re-establish their balance in this way, *have no dependence whatever on their seat*. This at once becomes apparent if you place them on horses not thus artificially broken in, or in situations such as happen in war, where the artificial training of the horse gives way to fear.

All experienced cavalry officers will tell us that the most docile and best-tempered horses are difficult to manage in battle. They sometimes go mad with excitement, and they then prove the most dangerous enemy the horseman has to contend against.* When nature thus becomes more powerful than teaching, when the horse in his fright forgets his education, and nature resumes its sway, then is the artificial horseman lost. Balanced on his fork, it is of no use pulling at his horse if he tries to bolt; for, with legs and stirrups behind him, the rider, at the slightest pull, falls forward, and has the greatest difficulty in keeping his balance. At the battle of Minden two entire French regiments were nearly destroyed by the horses taking fright and bolting in a charge. The men fell off and were trampled to death.

Without being ill-natured, I might relate many riding anecdotes which would amuse Englishmen; but I only mention the following to show how different are the ideas of foreigners to our own. They are scarcely necessary as showing the force of artificial training and inveterate habit.

Two foreign cavalry officers would not mount horses of the Royal Artillery offered to them for a review at Woolwich. I asked the reason; the answer was,— "These English horses are not broken in like ours, and might run away with us."

An officer in the foreign regiment I once served in took his horse over a low fence day. All expressed their admiration of him as a horseman, because he had actually taken the leap in an English saddle, which is supposed to be very difficult to sit in, compared to the military kit in which they are accustomed to ride.

I was once showing some foreign officers an English sporting print, in which the rider had his hand in his breeches-pocket, and a glass in his

---

* A disobedient servant and a disobedient soldier are equally useless; but a disobedient horse is not only useless, but often very traitorous and dangerous.—Xenophon, *On Horsemanship* [3.6].

eye, whilst his horse was clearing a fence. They asked me what it meant; had it any political meaning, or was it a caricature? I said No, it merely represented someone following the hounds. They all burst out laughing, saying,—"As if we are such fools as to believe that any man ever took a jump like that with a hand in his pocket! No, no! Englishmen may be cool fellows, but none of them can do that."

The difference between a school rider and a real horseman is this: the first depends upon guiding and managing his horse for maintaining his seat; the second, or real horseman, depends upon his seat for controlling and guiding his horse.

At a "trot" the school-rider, instead of slightly rising to the action of the horse, bumps up and down, falling heavily on the horse's loins, and hanging on the reins to prevent the animal's slipping from under him, whilst he is thrown up from his seat.

Foreign horses have little action compared to ours, and with them it may be endurable; but an English dragoon in marching order, trotting ten miles in this way on a powerful high-actioned English horse, is almost sure to sore his horse's back and his own seat. He wears out his constitution; for the strongest cheated man feels the effects of it.* He tires his horse more in those ten miles than any one else would do in fifty: he shakes his kit to pieces, and wears out his overalls.

Now, let me ask who can explain the advantages of this method of riding?

Foreigners will tell you it is necessary in order to collect and keep the horse in hand, as well as for parade purposes. Surely there can be nothing more distressing to the eye of a horseman than to see men holding on to the reins, and bumping up and down in the saddle; which, instead of collecting the horse and keeping him in hand, is, on the contrary, the cause of much unsteadiness in the ranks; for the unsteady seat alone is enough to excite a spirited horse, and the constant pulling at his mouth renders it in time callous.

The only two instances in which the method may be used to advantage are, when teaching the recruit to ride without stirrups, and when

*Our officers look upon their military seat with the lumping as part of their equipments, put it on when they fall in on parade, but wisely discard it at all other times.

working a young horse up to the hand on a mouthing-bit in the riding-school.

When cantering, the foreign school-rider never allows his horse to go straight, hut has him, in school parlance, "placed:" which signifies, when cantering to the right, the horse's head is bent to the right, and his haunches are brought to the same side by applying the left leg; when cantering to the left, it is the reverse: thus either way the horse is made to travel on the curve, with his head and tail drawn towards each other in an unnatural position. At a walk it is the same: few horses broken in on the system will walk with fore and hind feet on a straight line. Fore and hind feet move on parallel lines, the haunches being twisted to one side or the other; and the amusing part of it is, that the more perfect the horse is in his school education, the more palpable is his style of travelling on two roads at once.

The advantage of this *they* suppose to be, that the horse is always ready to turn: but why not let the horse go straight till he is wanted to turn? What would happen in an advance in line at a gallop if all the horses carried their heads and tails on one side?

A cavalry soldier should always ride straight to his point, and know how to "place" his horse, but never do so except in turning, or in striking off at a canter to either hand.

Instead of copying this seat and system from the foreign riding-schools, why not take example from our bold cross-country riders, adding to the instruction of our dragoons that skill in the breaking in and management of their horses with hand and leg, which will render them formidable singly; and that knowledge of riding pace which is so necessary to insure the steady working together of bodies of horse.

Give the man a roomy saddle, and make him sit close to his horse's back. Without drawing back the thigh, let the leg be supported by the stirrup in a natural position; the nearer the whole of the leg is brought to the horse the better, so long as the foot is not bent below the ankle-joint. Both man and horse will immediately feel the immense benefit of this return to national, natural practice; and, even without the adoption of any other changes, I feel assured that when next called into action our cavalry will play a distinguished and decisive part.

Our cavalry now is wanting in its most essential qualification—"riding." It is not sufficient that a dragoon can sit his horse; he should be completely master over him, so as to control and direct him at the slowest or fastest pace with equal ease; he should know how to quiet and subdue the hot-tempered, and put life and action into the sluggish horse.

If cavalry fought only in close bodies, if it acted like a machine, all required would be to discharge it at the mark like a projectile. Then, if the soldier could direct his horse anyhow to the right or left, move forward, and halt when ordered, it would suffice. But charges resolve themselves into mêlées, the dragoon is constantly exposed to the chances of single combat, and the unfortunate fellow who cannot manage his horse is lost.

Our soldiers are never taught to turn their horses quickly or make half pirouettes with them; which is of all things the most necessary in a fight. The reason it is not attempted is, that the advocates of the old system suppose that it requires years to teach a horse to pirouette, and they will not believe that, by the new system, horses are brought to do it, both on the fore and hind legs, in a very few lessons.

A fight on horseback is like a fencing match, in which the skilful horseman always presents his right side (which is under cover of his sword) to his adversary, and seeks to gain his weak side, the left one. Here all depends on horsemanship.

How is it then that so important a branch of military instruction has been hitherto attended with such poor results?

The fact is, they go on wrong principles. The instructors are not in fault, as they can do nothing with the system they are obliged to work on.

Now and then a cavalry officer is found that has broken away from the system so far as to train his own charger in a perfect manner, and to make himself a first-rate fencer and swordsman on horseback.

The Mysorean cavalry of Hyder Ali and Tippoo abounded in clever horsemen and first-rate swordsmen, who used their sharp weapons even with more effect than that with which the Sikhs have since been found to wield their tulwars. They frequently challenged our dragoons to single combat, and they generally had the advantage over them in the duels. But there was an officer riding with our troopers who had trained himself and his steed, and who could always give a good account of the best of them. This was Major Dallas—afterwards Lieutenant-General Sir Thomas Dallas—a cavalry hero and *model*—a sort of English Murat. Like that dashing Frenchman he was remarkable for his horsemanship and swordsmanship, for the strength, symmetry, and beauty of his person, for his daring courage, and for his love of hand-to-hand combats. He was sometimes seen to cut down three or four of the Mysorean champions, the one after the other, on the same day. He signalised himself, in the view of admiring armies, by many daring feats throughout the wars of Coote, Medows, Cornwallis, and Harris; and left a name that will be long remembered in India. We have many officers—ay, and men too—as brave as Dallas, and quite capable of doing all that he did, if they could only be taken out of the shackles of a bad system, and be properly trained to the use of proper swords.

We have had a variety of absurd systems in Europe within the last three centuries, and each of them, while it lasted, was productive of great mischief. Yet every one of them bad its bigots and enthusiasts, who looked upon any proposed variation or change as a shocking heresy. A master of the art, the celebrated Grisone of Naples, who was called the regenerator of the art of horsemanship in Europe, solemnly laid down the following instructions for his pupils:—"In breaking in young horses put them into a circular pit; be very severe with those that are sensitive and of high courage; beat them between the ears with a stick!!" etc. We now laugh at his pit and repudiate his stick; but both pit and stick had their reign, as other absurdities have had or still have. Grisone's followers, Pluvinel, Newcastle, La Gueriniere, Montfaucon, and others, substituted the cavesson, the longe, and the whip. They tied their horses to the stake (the pillars), and beat them to make them raise their fore legs, etc. I defy any one to find out, from their long rambling books on equitation, where to begin, how to proceed, or how to overcome by degrees each difficulty offered by the horse; and these difficulties, be it observed, arise regularly and in succession with every horse submitted to training. The question is, how to break in a number of horses, and upon what system to conquer these difficulties *one* by *one*? The old pedants could have given no answer to this question, nor am I aware that it can be answered by modern practitioners, or system-makers, or followers of existing systems or regulations.

It seems to me that the only man who has entered fully on the subject, and pointed out clearly how to attack: each point in succession, in order to gain the mastery over the horse, is Monsieur Baucher. But he has departed from the beaten track, has *disowned* the old system, and therefore his whole tribe have turned upon him.

I do not assert that M. Baucher's system is faultless. I practised it for years, applied it to many hundred horses, and was myself obliged to make some trifling alterations to adapt it to the use of cavalry soldiers. It may require further alterations to make it perfect; but what I assert is, that the system is the right *one*: it is founded on *reason* and *common sense*, not on immemorial custom and prejudice. So convinced was I of this by experience, that I wrote out and published the lessons as I had carried them out with hundreds of remount horses, to assist those who might be at a loss how to proceed when young horses join the regiments. Without system, and a good system, it is impossible to make good troopers. At present we have none.

The continual working at the horses' mouths now practised in the service, the attempt to throw the horses on their haunches by strength of arm, sawing the snaffle from side to side, teaches them to lean heavily on the hand, ruins their houghs and mouths, and wears them out before their time.

The attempting to work shoulder in, and passage, before the horse has been taught to obey the pressure of the leg, is simply absurd. It rouses their temper and makes them restive.

To rein a horse back before the head is brought home, and the animal has learnt to obey the leg, is equally absurd; for the horse with his nose stuck out can only be backed by force or by striking his fore-legs with a whip, a common practice in the riding-schools: he then steps to one side or the other; you cannot keep his haunches in the straight line, unless he has been taught to obey the pressure of the leg: the end of this generally is, that the horse gets his hind legs under him, is pulled back upon them, and the whole weight of man and beast is thrown on his houghs: the rider pulls again to make him go back, the poor beast cannot do it, no earthly power could move his hind legs, and in self-defence, and to escape from pain, the horse rears, the natural result of trying to back him by sheer strength of arm, and before he was prepared to yield to hand and leg.

The troop-horses that go through this rude treatment join the ranks with their action cramped and spoiled; they are seldom free from blemish, and their capability of long service has been greatly impaired. Mount any troop-horse, and you will find him hard-mouthed and stiff-necked. Few, if any of them, rein up or yield to the hand; they are all down on the forehand, and so accustomed to be held fast by the head, that, if you yield the rein to them in the least degree, they go off into a gallop at once, and then both hands of their riders are scarcely sufficient to stop them.

The system I have proposed rests on a few simple principles.*

I.—The horse is gently used, the progress is gradual but certain.

II.—For a few days he is ridden on the snaffle with a loose rein, at a walk and a trot.

III.—He is then bitted, and a few simple lessons teach him to yield to the feeling of the rein and the pressure of the leg.

IV.—Next he is collected and got in hand, not by pulling and sawing at his mouth, but by gradually pressing him with the leg till he raises himself off the bit and gathers himself up at a walk, when he can be collected and put together to any extent required, by the judicious use of the spur. As all this is done at the halt or at a walk, the horse undergoes no fatigue.

V.—Reining back then perfects the horse in the use of his limbs and in unqualified obedience to the rider's hand and leg. This once attained, a few lessons will teach the animal to canter, change leg, passage, and pirouette, and the horse becomes a perfect charger in a very short time, without having in any way suffered from his breaking—indeed, without

*The Training of Remount Horses, a New System. London: Parker, Furnivall, and Parker, 1852.

having been once tired or overworked during the whole of his education; and from his mouth having been gently dealt with, it remains fresh
and good, instead of being hard and callous.

I may speak confidently of these results, as I have myself obtained
them from the system with horses of various descriptions and breeds—
Arab, Cape, Persian, Australian, and the country-bred horse of India—
the last the least tractable of any. I feel assured that, with patience and
with a firm determination never to attempt to do too much at one time,
any cavalry officer may command the same, if not a greater degree of
success, with our English-bred horses.

Trained, ridden, and saddled in the way I have explained, the horse
will carry the weight well, the man will be less liable to be pulled off in
the ranks, his hand is likely to be light because he will not need its assistance to keep his seat, and, when required to exert his strength to manage his horse, his good seat will enable him to do so.

He will not fatigue his horse so much, and is less likely to sore his
back on a march; he will sit him easily over any obstacle, and make a formidable use of his sword, if, instead of standing up in his stirrups, he sits
close, carrying the weight and power of his horse into each blow, for in
this lies the great secret;—a child acting thus, in concert with his horse
as one body, will hit harder than a giant balanced on his fork.

It is of the first necessity that a soldier's horse shall obey the pressure
of the leg, otherwise he cannot be made to close up in the ranks or turn
quickly: but it is a great mistake to suppose that this cannot be done
without screwing back a man's legs and bringing them down almost
under the horse's stifle.

The habit of yielding to the pressure of the rider's leg is acquired by
the horse through teaching, and he will readily learn to yield to that
pressure wherever it be systematically applied. All practical riders, the
Cossacks, the Circassians, all Eastern nations, our own people—a
nation of horsemen than whom none more bold and clever,— all ride
in a short seat, and keep their own legs out of the way of the horse.

The Circassians are unsurpassed in the management of their war-
horses and arms, and so proud of their skill, that, whereas most nations
show wounds received in action as honourable scars, the Circassians

hide them as silent witnesses of their awkwardness and want of address in single combat.

At the Russian reviews in 1852 I saw a few sheets of paper placed on the ground opposite the Emperor: he gave a signal to some of the Cossacks and Circassians formed in line a few hundred yards off.

Down they all came at speed racing with one another: the first up fired at the marks either with pistol or carbine; the sheets of paper flew up in pieces: those who followed fired into the fragments that were at hand, blowing them to atoms.

The object of all preparatory drill should be to bring the dragoon to manage his horse thus at speed, and use sword and carbine at that pace—to teach him to reach objects on the ground with his sword, else an infantry soldier would only have to throw himself flat on his face, and when the cavalry had passed get up and shoot them;—a manoeuvre not seldom practised by old soldiers in war.

The Saxons, under Marshal Schulemburg, lay down to avoid the charges of the Swedish dragoons under Charles XII, during their famous retreat through Poland.

The Russian infantry, at the battle of the Trebia in 1799, were charged by the French cavalry when in line: they fired during the advance to the last moment, lay down, and, letting the French horse pass over them, got up and gave them a volley that emptied many a saddle.

A troop, by taking open order, should be able to go across country, for any distance required, at a rattling gallop, closing to their leader to charge when the signal is given.

All fighting with cavalry is generally either done at speed, or you advance at speed to get at your enemy.

In a pursuit of cavalry speed is the only pace at which you can catch the foe and destroy him.

To attain these ends we must have a great deal more out-of-door work. The soldiers ought to practise their various exercises in the open field. In making use of heads and posts we ought to scatter them in a field and allow the men to ride at them independently and chiefly at speed, in order to teach them to measure their distances with the sword, and deliver their cuts and points in proper time. As we use the posts in the riding-house, the man is guided by the walls, and learns nothing but to deliver the cut or point required, whereas the difficulty always is to measure distance at rapid closing; and this should be taught in the manner I have indicated. The Greeks were accustomed to train their cavalry to across-country work, to hunting and leaping. Where there were no animals to pursue they threw out a mounted trooper, and then sent another trooper to give him chase. The foremost man galloped through all sorts of places, frequently turning about, with his lance or spear presented to his pursuer: the pursuer carried javelins blunted and a spear of the same description; and whenever he came within javelin-throw he hurled one of his blunted weapons at the person retreating, and whenever he came within spear-reach he charged at him with the spear.* Here we see the counterpart of the Turkish game of the Djereed, the frequent practice of which tended to make the old Turkish irregulars such excellent cavalry. Since the days of European reform the old Turkish game is never practised!

Warnery says, "For a soldier to be really a light horseman he must be able to turn his horse quick and short when in full speed, to raise up and catch anything from the ground.

"Everything should be taught the recruit which may be requisite on actual service.

"He ought to be able to turn his horse suddenly upon his haunches, to run at the ring with the sword instead of the lance, which very much supples the horse and forms the trooper to dexterity and firmness in his seat.†

"As soon as the squadron is mounted the troopers are practised to leap ditches, enclosures, poles put across for that purpose, etc. At other times two troopers run together full speed, trying to get before and carry off each other's hats: they are practised to swim their horses across rivers, to manoeuvre in broken and intersected ground, etc.

"There are targets to be fired at by the troopers with their pistols, walking, trotting, and at full speed."

*Xenophon, *On Horsemanship*.
†Vide *The Training of Remount Horses*.

An English dragoon never rides at speed once during the whole of his drilling and training, nor ever afterwards, except when the squadron to which he belongs is ordered to charge; and then they cry out about English cavalry getting out of hand! Let any one think of the first time he rode a horse at full speed, and remember how helpless he felt!

I have heard it said that English horses are not adapted, like the Arab and other horses of Eastern breed, to skirmishing, to pulling up from speed, and turning quickly. The better the horse the more adapted to all feats of agility and strength. No horse can compare with the English,—no horse is more easily broken in to anything and everything,—and there is no quality in which the English horse does not excel, no performance in which he cannot beat all competition.

In teaching the trooper to ride I would make his first lessons almost as easy and simple as those given the horse.

When first put on horseback devote a few lessons to making his limbs supple, in the same way as you begin your drill on foot with extension motions. Show him how to close the thigh and leg to the saddle, and then work the leg back, forwards, up, and down.

Without stirrups, make him swing a weight round in a circle from the shoulder as a centre: the other hand placed on the thigh, thumb to the rear, change the weight to the opposite hand, and repeat the same.*

Placing one hand on the horse's mane, make him lean down to each side in succession till he reaches to within a short distance of the ground.

Vaulting on to the horse, make him place the left hand high up on the mane, the right hand swung back, but in the jump brought to the pommel of the saddle.

Off again in the same way.

Never when mounting with stirrups let him place the right hand on the cantle, for with an unsteady horse he cannot let go his hold to bring the leg over, and may be thus thrown, whereas, by accustoming the man to put his right hand forward on the pommel, the saddle is always open to receive him.

*Baucher.

These exercises give the man a firm hold with his legs on the horse, and teach him to move his limbs without quitting his seat

Then take him in the circle in the longe, and, by walking and trotting alternately, teach him the necessity of leaning with the body to the side the horse is turning to — *this is the necessary balance!* Then put him with others and give him plenty of trotting, to shake him into his seat. By degrees teach him how to use the reins, then the leg. Then put him through the lessons laid down for remount drill, beginning at the first lesson bitted,* and going regularly through the course, explaining to him the object of each lesson as you go on. In three or four months the soldier of common abilities will be ready for his further education at squad drill (when he must be taught to go across country).

He will learn to ride and to break in a horse at the same time—a great object to attain; for that soldier will be fit the next season to take a remount horse, and by pursuing this system, all dragoons being rendered equally able to do so, they could, on an emergency, prepare any number of horses for the field.

I would insist particularly on the propriety and necessity of explaining everything to the man as he passes from step to step in his instruction. School-boys and soldiers have been too long taught by rote, or without any proper endeavour to make them understand the use and object of what they are set to learn. Whatever the horse may be, the poor soldier is, at least, a rational being, and no good soldier will ever be made without awakening his intellect and reasoning faculties.

Write up in golden letters—or in letters distinguishable, and easy to read—in every riding-school, and in every stable: "*Horses are taught not by harshness but by gentleness.*" Where the officers are classical, the golden rule may be given in Xenophon's Greek, as well as in English.

The ancient Greeks had not only beautiful horses (originally imported from the East), but they had also great skill in training and using them as well for the saddle as for the racing-car. Although that short treatise is more than two thousand two hundred years old, there are excellent lessons in Xenophon, *On Horsemanship*.

*Vide *The Training of Remount Horses*

"In treating a horse," says the accomplished Greek, "this precept and practice will be found best—Never ill-use him through anger. For anger frequently excites to such rash and inconsiderate deeds, that they must be followed by repentance.

"When a horse sees anything suspicious, and does not wish to approach it, he should be made to see that there is nothing hurtful or fearful in it, more especially if he be a high-mettled horse: and if this cannot be done otherwise, the horseman himself must touch the object exciting terror, and lead the horse gently up to it

"Those who force horses forward with blows inspire them with still more terror. For, when they suffer punishment in such a situation, the horses fancy that the suspected object is the cause of it"

The whole of this treatise of Xenophon will well repay any one the trouble of an attentive perusal. This veteran soldier, historian, philosopher, and most elegant writer, evidently loved the horse with enthusiasm. On opening the essay, he says, "As it has fallen to my lot, from long practice, to have become experienced in horsemanship, so do I wish to point out to my younger friends how I think they can use their horses most properly."

7

ELEMENTARY DRILL

THE POWER OF CAVALRY IN the field depends upon the individual efficiency of the horsemen composing it; infinite care should, therefore, be bestowed on their training and teaching.

Riding, and the use of the sword and rifle, are points which should be brought as near to perfection as possible in the education of the dragoon.

The field movements depend upon the officers; the soldier soon learns to act his part in them.

The steady working of large bodies of cavalry depends entirely on the steady leading of the officers, and that again depends on their thorough knowledge of riding pace, the soldier naturally conforming to it.

After the men have been taught to ride in the riding-school, they must learn to ride in the field, and to handle their horses properly in the ranks.

To do this they begin work at open files; the instructor can then watch the riding of each individual and teach him his work thoroughly.

A squad, working thus at open files, should not exceed twenty men, and they should practise the following work.

To dress up singly and by ranks to three or four files, which have been moved up from either flank for that purpose, at a walk, trot, and gallop; the distances to be varied, and the proper pace to be strictly enforced; and no hurrying allowed at any cavalry drill.

To pass through the intervals by alternate files at a walk, trot, and gallop.

Right files march, halt.

Left files march through the intervals between the right files, and so on. Then, facing the files towards each other, repeat the same, to accustom the horses to pass through opposing lines. When practising at a gallop in all these drills, give the word, "*on the right leg*," or "*on the left leg*," "*gallop march*," according to the hand to which the dressing is; thus you keep up the breaking of the horses.

"Sit at ease!"

Files in succession advance, and engage on the right and left rein.

Files from the opposite flanks ride forward at a walk, then, turning towards each other, strike off at a gallop on the right leg, the sword at the engage; when abreast they come to the right guard, circle round each other, both simultaneously "*circle and change*" to the left leg, and engage in the left rein with the hanging guard; they file outwards, and the next two files follow.

"Advance in line!" The men should be taken over broken ground, no hurrying allowed, and the pace always steadily adhered to.

"Wheeling at open files," etc.

Part II. Working at close files and with two ranks. —The men are told on by threes; no other telling off is necessary.

"Mounting and dismounting." Instead of having different ways of doing it, as in the present system, it is best done at all times as laid down in H. M. Regulations (p. 143) for dismounted service.

In the formation by threes and files I would suggest the following alterations:—

"Advance by threes from the right:" the leading threes should move up thus:—

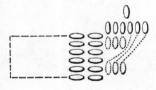

"Advance by sections from the right."

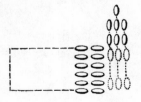

Every cavalry soldier should move in the column at the same moment when the word "March" is given.

This is impossible according to our manner of advancing by threes from the right.

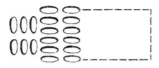

No man in the squadron can move till the leading sections have cleared the front of the following sections of threes.

The leading sections of threes alone move at the word "March," and each man in the column follows when he *thinks* it is time, or when he finds room to do so.

"Front form." The troop having gone threes right, they are to form to the front on the move.

No check of the pace should be laid down for the rear-rank sections, which always reacts on the column. Rather let the front-rank sections increase their pace, the leading one inclining to the right in front of the rear-rank section, and then dropping into the pace of the column. Its rear-rank section having continued to move steadily on, and the pace of the column having been kept up, the whole formation would flow steadily on with the stream.

Increasing and diminishing the front.

Advancing and retiring in line.

Inclining, passaging, and reining back.

"Dressing." Always make the cavalry soldier dress up, never back. Thus he learns to halt in time, in approaching the line he is to dress on.

Filing from the right of threes to the front, performing sword exercise on the move, at a walk, and gallop.

Ranking past by single files at a walk, trot, canter, and at full speed.

This is the best of all practice for cavalry soldiers; it keeps up the breaking of the horses, inspires emulation, brings each individual under the notice of his officer, and makes the men skilful in the use of their arms and the management of their horses.

On the caution, "Rank past by single files from the right," the troop-leader places himself in front of the officer on the right. The officer from the left flank rides down the front of the troop, and halts when his horse's head is in a line with the file on the right; his duty is to make the men close well up to the point from which they rank off.

The troop-marker rides on at a gallop from the right, to take up a position for the men to file to, and the troop serrefile places himself at ten yards from the right flank facing the passing line towards the troop.

Ranking past at a walk is done at one yard distance from head to croup; at a trot and a canter at three horses' lengths; and at full speed the distance is thirty yards.

The great difficulty in ranking past at a trot and a canter is to get the files sufficiently closed up to the passing line in order to feed the ranking off; and for this reason they should rank off first a front-rank, then a rear-rank man; thus both ranks close to the passing line and the men follow in quick succession; but if the whole of the front rank files off, followed by the whole of the rear, they come singly to the passing line instead of two at a time, and have to leave their places in the ranks at a trot or a canter; and as this latter pace would entail a change of leg in coming on to the passing line, the men would lose their distance, and the horses, instead of cantering to the right (when cantering past from the right), would most of them be on the left leg, as they would be obliged to turn to the left to get into the alignement.

On the caution to rank past at "speed," the front rank draw swords, the rear rank draw or unsling carbines.

A third marker rides at twenty yards beyond the first one, and thirty yards from the flank of the troop, on the passing line.

The men walk their horses the first ten yards to No. 1 marker, then strike off their horses at a canter on the right leg. At the second marker they let out their horses to full speed, bringing the swords to the engage; then perform the pursuing practice, endeavouring to reach the ground with their swords.

They must be told to give the horse his head in order that he may run straight, which he will do if they do not sway him from the line on which he is running, by standing in their stirrups and balancing themselves to one side and the other.

They should sit well down in the saddle, their legs well closed to the horse, and neither seat nor legs should move; the upper part of the body alone moves from side to side, and leans over to enable the man to reach an enemy on the ground.

The rear-rank men make use of their rifles and fire with blank cartridge at a sheet of paper placed on the ground opposite the reviewing officer.

An officer who sees a regiment of cavalry rank past at a walk, trot, and canter to both hands, as well as at speed, can at once tell what order that regiment is in.

He can judge of the bridling and saddling,

The state of the appointments,

The men's seats on horseback,

Their riding and the management of their arms.

The condition and breaking of their horses;

And nothing can escape him, for each individual in the regiment is brought under his notice.*

FORMATION OF THE TROOP

The troop is the *unit*, of which a certain number form the regiment.

It is perfect in itself, and requires no further drill but its own to enable it to take its place in a line or column and do its part, provided the officers are acquainted with the field movements.

The troop should be drilled by its officers.

The two divisions composing the troop form in rank entire, according to the size-roll, the tallest men and horses being on that flank which will be the inward one in the troop.

The officer commanding the division numbers it off from its inward flank, tells off the rear rank, consisting of the smallest men and horses, and forms two deep by filing or reining back and passaging. If the numbers are uneven, he leaves the last man but one on the outward flank of the front rank uncovered.

---

*On the 2nd of August, 1852, I saw a squadron of Saxon dragoons (120 horses) rank past at a trot without one horse breaking from it, and rank past at a canter to the right without one horse cantering false or disunited.

He then places non-commissioned officers on the flanks of both front and rear ranks: all others in the ranks.

The two divisions then close in and form the troop. Divisions having been equalised, the troop is told off by threes from the centre, no other telling off being necessary.

DISTRIBUTION OF OFFICERS AND NON-COMMISSIONED OFFICERS

One half a horse's length in front of the centre (troop-leader).

One officer on each flank dressing by the front rank.

Troop sergeant-major in rear of the centre (serrefile).

Troop-marker on his left.

Two trumpeters on the flanks of the rear ranks, covering the officers.

At open order the officers on the flanks move straight out and dress on the troop-leader. Trumpeters take their places.

The troop at drill should be practiced at—

Ranking past by single files, at a walk, trot, canter, to both hands, from the right or left; as also at full speed doing the pursuing practice;

Advancing by files, from the right, of threes, and performing the sword exercise at a canter;

All formations by threes and files—wheeling, increasing, and diminishing the front;

Dismounted service;

Skirmishing;

Advancing in line and charging;

Pursuing; and

Going across country in line.

If it be objected that I give the horses a deal of work, my answer is, that I have saved them a deal of work and much harsh treatment in the riding-school. My chargers are not worn out, but fresh, vigorous, and full of work. They can go the pace, and ought to be made to go it. They must be brought frequently into the field. Most foreign cavalry, when the gallop sounds, instead of increasing their rate of speed from the trot, actually decrease it, for they ride the collected short canter which they are taught in the riding-school, and only gallop when the charge sounds.

Matters are not quite so bad with us, but we shall not get a proper charge out of our young soldiers if they are never taught to ride at speed until brought into action.

8

On Intervals

"Il taut donner peu de front ut peu de profondeur aux escadrons, pour qu'ils aient de la célérité et de l'ensemble dans la charge. . . . On est souvent redevable de la victoire à de très petites divisions de l'armée, qui saisissent l'instant favorable à la charge. "Moins le front aura d'étendue, moins le désordre sera grand et fréquent."—Mottin de la Balme

IT IS CURIOUS WITH WHAT JEALOUSY intervals have been admitted in cavalry, yet many charges have failed in consequence of the confusion and disorder occasioned by the want of them. Horses and men exhausted by the pressure against each other in an advance can effect nothing when they join issue with the enemy.

The line, once on the move, cannot alter its direction, and, if overthrown, cannot get to the rear without carrying its second line along with it.

Mottin de la Balme, a very distinguished French cavalry officer, was himself in a charge of horse at the battle of Minden, and describes it as follows:—

"A corps of English infantry having dispersed the cavalry in its front by its steady fire, the corps of gendarmerie and carabineers received orders to charge. They advanced in line at a gallop: at first the centre was heavily pressed upon by the wings, then the pressure rolled back to the flanks, particularly to the right one.

"The infantry waited till we were close upon them, then opened a fire from the centre towards the flanks. The horses made desperate efforts to break away outwards and avoid this fire.

"The pressure became so great that men and horses upset each other and rolled about in helpless confusion; few were killed by gun-shot wounds, but, with the exception of about ten men in each squadron, they were all torn off their horses' back, trampled to death, or their limbs broken. The few that remained mounted were carried some right through the enemy's ranks, others to the rear or off the field.

"Had the advance been made by alternate squadrons we should have had plenty of room, the advance would have been made with speed and impetuosity, the horses could not have broken away to the right and left, and the English infantry must have been ridden over!"

The absolute necessity of intervals was at last acknowledged and some few introduced, precaution being adopted at the same time to reinforce or close up these openings by placing mounted parties behind them.

In some services these intervals vary in extent from four to twelve yards between each squadron. The Austrians, however, put two squadrons together, which they call a division (nominally 300 strong), and leave an interval of twelve paces between those divisions.

I have ridden in these Austrian divisions, and the pressure of the horses was often so great as to lift me, with my horse, off the ground, occasioning great pain, and making one and all quite helpless. A few resolute horsemen dashing in on such a mass would throw it into utter confusion.

Where the ground is perfectly level, where there are no undulations, no natural obstacles, an advance of squadrons with small intervals can be made, though with difficulty, particularly under fire, when the horses crowd together from fear; but if the country is rough, broken, or intersected (which most countries are), it is impossible to advance quickly and preserve order.

When pressure takes place, it rolls on and on, increasing like a wave, till it runs itself out at the first open interval; but squadrons expand when they take flight, their intervals are thus closed, the waves meet, break the line, impede its advance, and throw it into irremediable confusion.

The vicious and fiery horses contribute not a little to this result. The pressure drives them mad; they throw themselves against each other, burst from the ranks, or press out the weak ones from the line, and many become so exhausted in the struggle as to be unable to keep up.

If in tactics you aim at the greatest precision, the consequence of the smallest mistake is not *less precision*, but utter confusion. Thus, to advance in line with small intervals and large squadrons is too much to expect under all circumstances; for what would create confusion when standing still is ten times worse when on the move. Wounded men go to the rear; wounded horses fall or break from the ranks, causing confusion and difficulty enough without stinting the men for room, winch makes order an impossibility, and detracts naturally from the speed and impulsive power of the cavalry.

If every fifty horsemen had an interval of twelve yards in the advance, their natural expansion would fill up one-half of the intervals; the remaining half would prevent their crowding, and give each body sufficient play to ensure its freedom of movement; thus they could charge over the most difficult ground close to their own centre (easily done by fifty men), and dash in close array against any line opposed to them.

A charge, even on good ground, is seldom executed by the whole line at once; the enemy is reached in succession by different points in the line more advanced than others. It is therefore of the greatest consequence that those detachments which reach the enemy *first* shall be compact, and go at him *as one man*, to burst through. It is easily understood that with fifty men this can be done; but it becomes almost an impossibility with one hundred and fifty or two hundred.

I should therefore form each troop with the tallest men and horses in the centre, keep it up to twenty-five files in the field, place the captain in front, the subalterns on each flank, and give them twelve yards interval in line.

All formations would then proceed easily and freely.

When wheeled into oblique échellon they would not overlap each other, nor require to incline to clear their front when forming line.

They could act independently when necessary; and when required to act in concert they could do so more quickly and with more energy, because, as they can both see and hear their leader at all times, their efforts would be concentrated round him.

They could always alter the direction of the advance, if it were suddenly found necessary to do so, without breaking into column.

And when in column, they could work with greater ease in consequence of the additional distance which the interval of twelve yards affords from the troop in front.

They would succeed against an infantry square when a large squadron might fail, because they could advance with greater speed. The officers on the flanks could prevent their opening out and riding round the corners. The chances of being hit would be lessened (both by speed and the smaller number), less confusion would ensue, and the men, conscious that their conduct must be observed by the officers, would do their duty.

Again, if four such troops of fifties fell in with an enemy's squadron of two hundred, whilst one or two troops attacked in front, the others would fall upon his flanks or rear; or during the time that the great unwieldy squadron was trying to form line, one troop of fifty, dashing in, would throw it into disorder: the other troops, led independently by their leaders, might then choose their own points of attack, charge, and overthrow it.

They would rally more quickly, and would do good service over country which would entirely disorganize large squadrons in an advance.

It is not necessary for cavalry to be numerous to achieve success, but bold, resolute, and rapid.

Kellerman retrieved the battle of Marengo with a few hundreds of such horsemen.

On the third day of the battle of Arcole, Napoleon detached Captain Herculc and fifty dragoons with orders to gain the left of the Austrian army, and charge it in flank whilst he attacked in front. The charge was gallantly executed, and contributed greatly to the success of the day.

If Napoleon considered fifty horsemen sufficient to attempt a charge against the *whole Austrian Army*, surely the same number of horsemen will be found sufficiently strong to take the place of a squadron in line.

9

FIELD MOVEMENTS

*"On doit chercher sans cesse, avec un soin scrupuleux, à simpli-
fier les exercices de la cavalerie, que tant d'innovateurs de nos
jours ont mal-àpropos compliqués. Dans ces vues, il faudra
nécessairement former et faire combattre cette troupe sur un
front pen étendu."*—Mottin de la Balme

THE ART OF MANOEUVRING CONSISTS in attacking your enemy at his
weak point, or falling on whilst he is in the act of deploying.

The field movements enable you to manoeuvre with large bodies of
men, and give you the means of forming them quickly on a given point
or line for the charge.

How many times might a successful charge have been executed, how
many times a victory gained, BUT that the column was either right in
front when, to suit the emergency, it ought to have been left in front, or
the reverse!—and on such things hangs the fate of battles.

What can be more preposterous than the position of a commanding
officer leading a column into action, obliged to weigh his chances and
calculate them versus right or left in *front*, or which is the pivot?

It must make an officer nervous, when told to advance, to think that
honour and renown, death or disgrace, may depend upon the choice he
makes between the right flank or the left one.

Dr Brack, in speaking of some field movements, not forming part of
the drill-book, says, "True they are not laid down in the regulations, but
they are necessary, because they are *simple*, and fulfil the grand requisite
in war for cavalry, that of being able to front promptly in every direction. By emergent and unexpected movements in action, the order of
the squadrons in the regiment is upset, and circumstances imperiously
require you to manoeuvre in this state under penalty of annihilation."

Almost every authority on cavalry warfare will tell you the same, and yet you refuse to practise in peace-time what in war becomes a necessity. "When," says De Brack, "by a sudden emergency, we are compelled to form a line in which the squadrons and troops no longer stand as they have been accustomed to, we are all abroad, and hesitation and danger ensue."

In France, Colonel Itier, of the 7th Chasseur à cheval, proposed a system by which you always worked and formed to the right: this would obviate the difficulty of choosing between the one and the other; and as it would be impossible to form, except in the one way laid down, there would be at least no hesitation; but it is like telling a man who has two good legs to use only one.

Colonel Itier worked by divisions, and inverted them in the squadron, and inverted the squadrons, or mixed them, in the brigades and divisions. He had a close and an open column of divisions. This is strange, for you never can close up divisions sufficiently to get your distance for wheeling into line. The Austrians, he it observed, call a column of divisions a close column, for this reason.

These pivot flanks and right or left in front have been a plague, a puzzle, and a cause of mischief, to all military men ever since the period of their first introduction. This, indisputably, being the case, is it not high time to see if something bettor cannot be substituted for them? I think that here, as in every other matter connected with the cavalry service, we shall find our best chances of success in a return to simplicity. What is easiest is generally best in the long run.

Tacticians lay down as law, that

The component parts of a column shall always follow each other, as

    a
    b
    c
    d

—and that, when brought into line, they shall stand thus:—

   d   *c*   *b*   *a*

For instance, the column *a*, *b*, *c*, *d*, wishes to form
line to the front: it can do so to the left of a, because b
comes on the left a, c to the left of b, d to the left of c
and they will thus stand as required by tacticians,
namely:—

a

b

d

$$\underline{d} \quad \underline{c} \quad \underline{b} \quad \underline{a}$$

But they cannot, must not, form to the right of a, not that there is
any natural impediment in the way, but because they would then stand
in line thus:—

$$\underline{a} \quad \underline{b} \quad \underline{c} \quad \underline{d}$$

Let us take an example.

The army A, drawn up in two lines, expects the enemy in its front.
Intelligence is brought of the enemy's columns, B, moving perpendicu-
larly to their left flank.

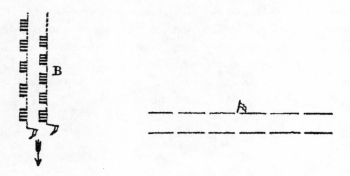

The army at A immediately marches off in columns to the left, but,
instead of hitting on the head of the enemy's columns at c, the enemy's
columns have marched so quickly, and gained so much ground to the
rear, that the A army strikes upon the rearmost divisions, which have
been brought opposite to them by the continued march of the B army.

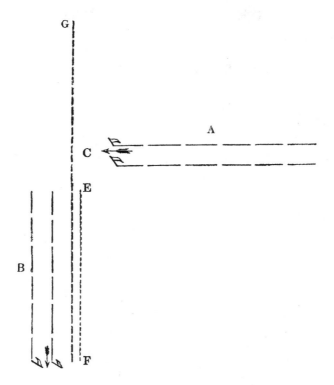

A army in column, left in front.

A wishes to engage B: to do so, he must form line opposite B between E and F. Unfortunately this is impossible: he is LEFT IN FRONT, and must bring his right to G: as he can only form line to the front between E and G, he cannot bring the enemy to action.

The tactician wants to measure his distances, to put every individual in his proper place, and to teach him what he has to do; some little thing goes wrong, and the whole force is thrown out; for on the exact fitting of all the parts depends the working of the system. The tactician is drilled and drilled till, when perfect, he can do nothing but what he is ordered; he loses all self reliance, and in an emergency cannot break from the trammels he is held in; but the less perfect tactician. would act on the spur of the moment.

The knowledge of tactics no more makes the general than the knowledge of the number of syllables required in verses makes the poet. Genius alone can make the poet and the general.

The real general, when going into battle, seizes on all the advantages of a position, on the openings given by the enemy, as if by inspiration; there is no hesitation, ways and means are never wanting.

If the troops he commands are capable of moving under fire, he places them by a simple movement within reach of the foe. He quietly waits the result; sending reinforcements where they are required, or occupying in time, and in strength, positions to cover his retreat should the day go against him.

Ziethen, who was born a general, whose intuitive perceptions always led him right in battle, never got on well at the complicated sham-fights held by Frederick at Potsdam to try his generals. Frederick, thinking that a little working up in tactics would do Ziethen good, sent for him, and examined him as to what he would do with his cavalry under certain circumstances. Ziethen answered very quietly— "Well, I don't know just now; when the circumstances happen in the field, and I see the enemy before me, *somehow* it will be sure to occur to me."

A most curious instance of the difficulties to which this system of pivot flanks leads, happened before the battle of Prague, and I choose it in preference to any other, because it completely puzzled *Frederick the Great*, himself the greatest tactician of the age.

The night before the battle of Prague, Field-Marshal Schwerin's army, formed in two lines, broke into column *right in front*; the artillery between the columns, the cavalry following the columns formed by the second line.

The king marched off *left in front* in two columns, the infantry in one, the cavalry in the other.

The king and the marshal met each other at an obtuse angle on the ground where they were to form: and General Winterfeldt marched in between them with his column *left in front*.

Marshal Schwerin's columns were destined to form the left of the army, which was to advance by its left against the enemy; but no one knew how to unravel the mass without great loss of time: at last the

Duke of Bevern, with great presence of mind, made the rear of each of the Marshal's columns take ground to the reverse flanks, and move forward followed in succession by each detachment in it.* Thus originated one of our field movements—"Rear of the column to the front."

But if an enemy came on suddenly, this manoeuvre like many more, could not be performed, and serious consequences might ensue.†

We will, therefore, have no pivot flanks. Troops, when in line or in column, shall be led by their centre, and during all formations by their inward flanks.

The troops (except on parade, when they fall in by seniority) take their place in the line of battle as opportunity offers or necessity requires. No inversion is possible.‡

The troop is the unit of any number of which the regiment may be formed. These troops file or move by threes or sections from either flank, and form at twelve paces interval from each other in line.

They are alike in every respect; it makes no difference where they stand in the regiment.

The troops often change from one wing to another during the movements; as, for instance, the regiment is in line composed of the following troops—*a*, *b*, *c*, *d*, *e*, *f*—and advances In double column from the centre, closes up, and deploy to the right; the troops will stand in line *c* *d* *b*, *e* *a* *f*. The left wing is now composed of the troops *c*, *d*, *b*, and the right wing of the troops, *e*, *a*, *f*. This makes no difference; the troops are all the same; and of these troops a certain number, not certain troops, form the wings.

Troops of different regiments cannot mix.

Regiments fall in by seniority on parade, but take their places, when manoeuvring, according to circumstances.

---

*Berenhorst.

†In the following field movements proposed, details are not given, for such details are matters for after consideration.

‡Fortunately the English cavalry have remained organised in troops, whilst all continental troops are in squadrons, which they bring to 200 horses and upwards in wartime. To be even with them we might call four troops a squadron, and two troops half a squadron. This would facilitate the working in the field particularly if all regiments had two such squadrons, or eight troops.

The commanding officer gives the word of command, and, as it is of great consequence to have as little noise as possible, in order to avoid confusion, the two majors or wing-leaders alone repeat that word of command, and ride at about fifty paces in front or on the flank of their wings.

Their trumpeters alone repeat the signals, except in the charge and rally, when all join in.

The adjutant rides with the colonel; the superintendence of the base is confided to him.

Markers take up the inward points, and, instead of dressing close up to them, the line is formed at some distance behind the markers.

Troop-leaders only give a word of command when it concerns their own troop.

The pivot flank is that which brings the cavalry quickest upon the foe.

The commanding officer is hampered with no conditions, and forms his line in the shortest and most simple manner to ensure success.

Paces.—From the halt the line breaks into column, or the column wheels into line at a walk or a trot.

The horses are only knocked about and excited by trying to do it at a gallop; the distance is so short that it is not done one instant sooner, but rather the reverse; because in doing it at a gallop they fly out, and have to close in or rein back.

When on the move all wheels are made at an increased pace.

The trot.—For manoeuvring the proper pace to use is the trot, at the rate of eight miles an hour.

The canter.—A distinct line should be drawn between the canter and the gallop: the former should be as collected and as short as possible. It is necessary for a man to be able to collect his horse to the very slowest pace without breaking into the trot, in order to handle him in single combat.

They should have plenty of practice at cantering to both hands not only in the riding-school but at drill, by ranking past by single files to either hand, doing the sword exercise in line at open files, etc.

The gallop should be at the rate of fourteen miles an hour, and seldom used except for forming line to the front from an open column; and in the advance, where the men must keep it up across country, without pulling at their horses or hurrying.

The great point is to get the officers to take up the pace at once, and keep to it steadily throughout: that is, in an advance at a trot or a gallop, the officers ought never to allow the rate of either to be increased as they go on, otherwise the trot would soon break into a gallop, and the gallop would end in the men going off as hard as their horses could run.

Cavalry that hurries is never ready when wanted, because it is "out of hand." All hurrying must therefore be checked at once, and at all times, for it is the only way to prevent it getting "out of hand:" a fault to which our cavalry has always had a tendency. In their charges men, and even officers, would often dash forward during the advance, causing the alignement to be lost, and breaking the power of the charge. When a regiment is advancing under fire, men and officers get excited, hurry forward, and at last, goaded on by the shots dropping among them, burst from all control, and gallop madly towards the enemy.

This is nothing more than *running away*; though it is running forward instead of back. It shows the same want of soldierlike endurance and fortitude under trying circumstances, and often leads to defeat and dishonour; for they all get out of hand, scatter, and are driven back with loss by the enemy's reserves.

Men and officers should therefore understand that to gallop forward because the enemy are in that direction is by no means a proof of valour, but often the reverse; and such conduct ought to be severely censured and checked on all occasions.

General Sohr, one of the most distinguished Prussian cavalry officers, was keenly alive to the necessity of preventing such attempts on the part of individuals, as the following extracts from his Life will show:—

"During the advance of the Prussian army into France, on the 3rd of February, 1814:—

The 1st Dragoons	4 squadrons
The Lithuanian Dragoons and mounted Riflemen	5 squadrons

The 3rd regiment of Hussars	5 squadrons
The Brandenburg Lancers	4 squadrons
The Mecklenburg Hussars	4 squadrons

Total 22 squadrons—

This force marched at six o'clock A. M. on the high road to Châlons; at seven they fell in with Sebastiani's French corps of cavalry near the village of La Chaussée, consisting of seven regiments of cavalry, chiefly cuirassiers and lancers, lately returned from Spain. One of the greatest cavalry engagements of the campaign now took place, called the battle of La Chaussée.

"General Katzler, seeing that the enemy (who were superior in numbers) were anxious to form this side of the village, gave orders to fall upon them whilst endeavouring to form line, and we were hurried forward to the attack, for the rumbling of the enemy's artillery was distinctly heard coming through the village, and our General would not give them the chance of getting into position.

"The necessity of haste and bad ground prevented all the regiments from coming into line and charging together. It therefore resolved itself into an attack in echellon, or in succession; but each regiment was led steadily to the charge by its commander.

"The charge was carried out with the greatest determination; each regiment pushed eagerly forward to close with the foe.

"We had the advantage. The Brandenburg regiment of hussars was lucky enough to break through the enemy's centre (composed of cuirassiers) and to make many prisoners, though they fought most gallantly. We captured four guns and two ammunition waggons in the act of preparing for action.

"The enemy, driven through the village of La Chaussée, tried to make a stand on the other side, but was overthrown and pursued some miles towards Pagny.

"The engagement lasted till dark. General Sohr, then in command of the Brandenburg Hussars, displayed great coolness and energy in the fight, and I cannot refrain from relating what took place, to prove how necessary it is in warfare to show determination at all times and under any circumstances in order to keep troops in hand.

"Whilst the regiment was advancing the rash behaviour of one of the officers had nearly caused the charge to fail.

"The gallop was sounded; Count v. d. S., a lieutenant who had lately joined from the Saxon cavalry, wishing to distinguish himself, brandished his sword on high, called to the men to follow him, and dashed forward at speed. The second squadron, which began to follow him, lost its place in line. Sohr immediately ordered the trot to be sounded, and then, waiting till the whole regiment was steady, he sounded the gallop and the charge, the only means by which he could hope to break through the enemy's line. When the engagement was over, Sohr called the officers together, and addressing himself to Lieutenant v. S., said, 'You proved to us this day that you have lots of pluck, and I honour those who have, but I am myself no "Hundsfott," and if you again forget yourself as you did this day, and dare to interfere with me in the leading of the regiment, I shall cut you down in front of the line.'

"The lesson was a severe one, but such examples seldom fail in their effect.

"Once more Sohr found it necessary to enforce steadiness in his squadron, but on no subsequent occasion was it necessary.

"To check the pursuit of the enemy, a strong rear-guard was pushed forward, at three o'clock A.M., from Weissenberg to Wurschen. The enemy was brought to a stand by the sudden opening of our artillery, and more than an hour elapsed before he received sufficient reinforcements to resume the offensive and drive in our rear-guard.

"At the pass of Rothkretcham, to the eastward of Weissenberg, is an arm of the Lobaur stream, the passage of which was hotly contested. Sohr's squadron was formed in the plain in front of the defile, and the enemy's troops, of all arms, were seen pressing forward on the neighbouring heights to the north and westward. When our rear-guard had effected its retreat through the defile, Sohr thought it high time to do the same, and gave the word 'Divisions, right about wheel—march!'

"The enemy was near at hand, and the divisions wheeled about hurriedly, and almost before the command was given.

"The experienced leader, who had an eye to the future behaviour of his squadron, immediately fronted the divisions again, and, placing himself at their head, said, 'I'll have you all cut down by the enemy rather than see you work unsteadily.'

"He faced the foe; not a sound was heard in the squadron; the enemy pressed on; and their artillery opened with round shot on the defile in his rear, cutting up the ground on both sides of the devoted band. The sudden fronting and bold attitude of Sohr's squadron fortunately led the enemy to suppose that supports were at hand, and they ordered a flank movement to turn his position. His situation became more critical every moment he remained, for the enemy's cavalry were now coming up— still not a movement was perceptible in the squadron till, turning his horse towards them, he gave the word in his peculiarly measured way:—

"'Divisions' . . . . 'right about wheel'. . . . ' Walk'. . . . ' March'. . . 'Threes right!' and in a voice of thunder he added, 'At speed!' 'March!' 'Ride as hard as you can!' The defile was passed almost together with the enemy. Never again did his squadron hurry. In the hour of danger the hussars looked with confidence to their leader and 'were. in hand.'"*

Field movements are of two sorts, namely, those which are made use of when close to and immediately before engaging the enemy, and those used when not in the immediate vicinity of danger.

The first, an open column of troops, ready to form in the shortest manner to front, rear, or flanks.

The head of each column under its own commanding officer.

Each separate body in the column under its own leader, and of any number of such separate bodies the column is composed.

Thus, if surprised, they act for themselves, and may still win the day.

All open columns should be formed of troops; the troops kept up to twenty-five files.

Columns of smaller front have many disadvantages which are not compensated for by the only real advantage they possess, namely, that of saving the horses some work in the wheeling, as they cover less ground than the column of troops.

In columns of divisions there is great difficulty in keeping the distances; you never can get them sufficiently closed up, even on a parade-

---

Aus dem Leben des K. Preussischen General-Lieutenant Frederick v. Sohr, by H. Beitzke, Major A. D. Published in Berlin, 1846.

ground, and when they are closed up sufficiently the rear rank of the preceding divisions is always in the way in wheeling into line.

Divisions have been used in preference to fours and threes when under fire, because no tellings off are required.

The column of divisions is simply moved off its ground to disengage the dead and wounded.

But it must be remembered that formerly troops were told off in four or five ways, and this by the sergeant riding down the ranks, pointing with his sword to each man, and first telling him whether he was a centre or left by threes, then a right or left by files, and so on: this was a tedious operation.

Men tell themselves off now, and that in an instant, particularly as the only necessary telling off is by threes; and old soldiers will go threes right and left without any telling off.

A column of divisions is of all columns the worst for going quickly across country; they are so close to each other that no obstacles can be avoided or cleared. It is like a close column. In dry weather the dust has no time to be carried away by the wind. It is most difficult to hear the words of command. A sudden check throws such a column into confusion, and, if attacked, the divisions, trying to form troops and then squadrons to be under their legitimate leaders, would detract greatly from the chances of success.

All these things considered, I should prefer the column of troops, and only use divisions when a troop, out skirmishing or otherwise detached, requires a support to its advanced parties.

The advantages of the column of troops are numerous; some of them are set forth in the chapter on Intervals, the others will become apparent as we proceed with our field movements.

All formations from the open column must be made towards the foe, that is, the hack of no single horseman in it should be turned on the enemy during the execution of the movement,

Should the enemy attack during a formation made towards him, no great harm is likely to ensue; the troops would meet him in succession, and charge whilst following out the line of formation.

The column forms line to either flank by wheeling into line, or forming in succession to either flank.

It forms to the front either right or left, as ordered, the commanding officer always placing himself and giving the word of command on that side to which the troops are to form; one note of the trumpet signifying "right," two notes "left," which signal should always precede the word of command.

Majors shift their flank with the commanding officer.

Formations to the rear are done by wheeling troops right or left about, and proceed as above.

In forming line from column a flank is brought up or thrown back by the commanding officer placing the base troop, and giving a caution, before the word "March," to bring up or keep back the flank forming.

When the head of a column has changed direction, line may be formed on the new alignement to either hand; the rear of the column either coming up in oblique échellon, or going threes outwards.

When not in the immediate vicinity of the enemy, line may be formed on any central point of the column, the adjutant marking the troop on which the formation is to take place; all the remaining troops wheel to the hand to which the formation is ordered, and those in front of the base are then put threes about; thus all possibility of a mistake is obviated.

When advancing against an enemy in the open, do so by double or single contiguous open columns, either of wings of regiments or of regiments; you thus avoid the disadvantage of a single column, which requires time to form line to the front.

A column of troops in passing a defile breaks into threes, and reforms on coming out; breaking into threes does not lengthen the column, and the interval of twelve yards will prevent any check being given, which would most likely occur in breaking into divisions, and again in breaking from divisions into threes.

It would be an immense advantage to have all these things done *one way*; of course when the passage is too narrow to admit of threes, the column must halt and break off by sections or files.

Double open columns, under the existing system, have the advantage over a single column that they extend over only half the ground, line is more quickly formed, and words of command more easily heard.

The objections are, that their greater extent of front makes them unwieldy, and they require more favourable ground to work on; they cannot show a front to the rear without first forming line to the front, and they cannot break into single column.

I propose that, in order to make this column useful, it can be formed, as it is now, from line or from a close column by simply opening out, or from contiguous open columns by closing two such columns to each other.

It forms single column by the troops on the left halting and inclining to the right, and re-forms double column in the same way by the alternate troops moving up to the left at an increased pace. They form contiguous open columns by opening out from each other. The double columns can thus suit their movements to any ground; for instance, an obstacle in front of a double column obliges them to open out to avoid it, the enemy suddenly appears, and they act as contiguous columns; if contiguous columns closed in to each other, for the same reason they are then ready to act as a double column.

A double column threatened suddenly in flank wheels the troops on that side into line; the other wing may wheel into line also, and follow as a support, form in succession, or move on till clear and form in échellon, &c.

It forms line to the front, or either flank; and by wheeling troops to the right or left about it ia ready to form to the rear in the same way as to the front.

Double columns are useful in passing defiles; but each regiment should form its own double column and follow in succession, for, by forming double column from the centre of a line of cavalry the regiments are mixed, and if attacked by the enemy before they have re-formed line it will go hard with them.

On issuing from a defile it may be found necessary, after the first regiment has formed, to bring the next following to form on its right or left, which could not be done if what had debouched of the column was composed of the half of a regiment from the right flank and of half a one from the left flank.

The object must always be to get into line quickly, since the "charge" follows; therefore always obviate the chances of losing time, for with the loss of time, the chances of success are always diminished.

In going into a defile, it is impossible to give orders as to how to form on the other side; this must depend upon the enemy's arrangements to receive you.

The formation must be made parallel to the line of attack; circumstances are not always favourable; the ground may be bad. The officer commanding has no time for consideration, but must form and charge to try and keep the débouché open for those following.

This can be done very well if regiment succeeds regiment in double column, the colonel leading, his majors on his right and left, each in front of his wing; thus each regiment is ready to act at once in any part of the field, on either flank: and in case of disaster each regiment being together is more easily rallied and re-formed.

To rally quickly is of vital importance, and for this reason it is a great mistake to assimilate the uniform of all regiments, for it adds greatly to the difficulty of rallying in presence of an enemy.

In some of the Prussian regiments each squadron has different facings, to facilitate their rallying in battle.

In the Danish service they distinguish the squadrons in the regiment by a badge worn on the breast.

But, whether it be done by colour of the jacket, by facings, by badges, or by any other method, I certainly consider it a primary necessity that our cavalry regiments should be markedly and strongly distinguishable, even at a distance, by their several uniforms, in order that the rally be not impeded.

Close columns are chiefly for the assembly of troops or reserves: to bring them in the neighbourhood of an enemy would be dangerous, for they cannot show a front to their flanks, and consequently cannot protect them. If the enemy attacked the head of a close column, and at the same time gained its flanks, the close column would be destroyed, for it could not develop its strength.

Thus, when approaching the enemy in an enclosed country where a surprise might be attempted, close columns must open out and form double or single open column.

The only occasions on which close columns may be used advantageously for fighting are when a regiment is surprised in the open and cut off from its supports by a superior force. It should then collect into a compact body, make a gap, and cut its way through.

Light cavalry regiments have acted successfully in close column against lines of heavy cavalry advancing to the charge, their object being to break the army and take the heavies singly in the mêlée: to do this whilst the adverse lines were advancing towards each other, the lights suddenly formed a close column on their centre at a gallop, and at once bore down on the centre of the line; thus avoiding the charge, breaking through the line of heavies, and throwing them into confusion, they spread out on their flanks and rear, and attacked them singly and with an advantage.*

Cavalry may be brought up in close column, if its flanks are protected and its deployment covered by a powerful artillery, but under no other circumstances.

The front of a close column should be composed of two troops in line, with the interval of twelve yards between them. Always keep the front of the leading squadron clear both in forming and deploying, except when it is necessary to form or deploy on a central squadron.

In forming close column to the right, all squadrons go in rear of the right one, and *vice versa* to the left.

In forming on a central squadron the right squadrons form in front, those on the left in rear of it, whatever the number of squadrons or regiments may be.

Now, let there be any number of regiments in close column, the officer commanding requires a line to the right.

"Deploy to the right" All go threes right, except the squadron at the head of the column, which gives the base. To the left, *vice versa.*

Perhaps the General wishes to deploy from a central point, bringing more squadrons to one flank than to the other: for instance, he has three regiments in one close column, and wishes to deploy two regiments to the right and one to the left.

*De Brack

" Deploy to the right on the second regiment."The leading squadron of the second regiment is the base, *therefore*, all in rear of it go threes right (the hand named).

The first regiment is in front of the base, and goes by threes to the opposite flank named for the deployment, and this is the rule, namely—

All squadrons in rear of the base go by threes to the hand named for the deployment.

All squadrons in front of the base go by threes to the hand not named.

If the General requires two regiments on the left, and one on the right,—

"Deploy to the left on the second regiment." Leading squadron, second regiment, is the base; all in rear of it "threes left" (the hand named).

The first regiment, which is in front of the base, " threes right" (the hand not named).

Thus the officer commanding requires no knowledge of how the columns were formed to unravel them at once in any direction.

When a close column, composed of more than one regiment, requires to front to the rear, it must be countermarched by ranks in each troop, because it would take up too much time and require too much space to open out such a column; but contiguous close columns of regiments are opened to double column distance, all simultaneously wheeled right about by troops, and then closed up again. Countermarching is one of those movements never to be employed when near an enemy.

When regiments in close column are made use of on the field, they should never be formed in one column, but in columns of regiments either contiguous, at deploying distances, in échellon or échiquier (alternate); thus they at once become available for any emergency by opening out to double-column distance.

Echellon and contiguous open column.—The combinations of which the échellon and contiguous open column are capable greatly simplify manoeuvring, and supersede the necessity of many field movements.

They should not be formed of less than wings of regiments; when smaller they do not unite the same advantages.

Changes of position by threes or oblique échellon of troops are objectionable, because the moment of formation is always a weak though an unavoidable one for cavalry: the more ground the formation extends over the greater the danger, and to avoid this you should bring your troops in columns close to the line of formation before giving the order to form.

Changes of position are best executed by a combination of the échellon and contiguous column.

Single, double, and close column can all be formed at once from either.

Changes of front should be executed on the same principle as laid down in Her Majesty's Regulations.

The advance in line.—This has generally been considered as simple enough; yet it has long appeared to me that the arrangements made for the direction of any advance in line are insufficient and faulty.

All manoeuvres are preparatory only to the advance; and when the line is formed and launched forward on the enemy, if he is suddenly found to be on the right or left front instead of straight before the advancing line, it is too late to alter the direction on the move, and to halt is dangerous; what with the difficulty of hearing words of command, particularly when under fire, and the impossibility of seeing what the base squadron is doing, the cavalry cannot easily be brought to bear straight down upon the foe.

The squadron leader often loses sight of the base altogether; it is buried in the ranks, and, being on the same line with him, he cannot distinguish by the position of their horses whether they are inclining towards him or from him.

If the base squadron changes the direction of the advance towards him, he first becomes aware of it by being run into and pushed out of his place in line.

If the base squadron changes direction away from him, he is left suddenly a long way behind.

The constant endeavour to watch the base makes him often increase the pace of the squadron unnecessarily, and obliges him to pull up as suddenly when he finds that he has overshot the mark; the constant looking sideways renders it difficult for him to lead the squadron on a straight line.

I think there ought to be a guiding power established independent of all words of command, and so arranged that the commanding officer might, through it, influence the movements and direction of the whole line at once.

This could be done effectually by a very simple arrangement.

The adjutant places himself 25 to 50 paces (according to the extent of the line) in front of the centre of the regiment; the two troop-markers of the centre troops gallop out and place themselves at five paces interval on his right and left, and raise their swords; forming a conspicuous group visible to all.

The adjutant (who rides with the colonel) knows what point he is to lead on. He is answerable for the pace.

Troops dress to their centre, and the two centre troops of the line keep exactly behind the base, conforming to their movements.

When the colonel wishes the direction altered he has only to ride up and tell the adjutant, who, with his flankers, at once circles into the new direction, all of them pointing to it with their swords until it has been taken up by the line.

The troops on the flank which is brought up increase their pace, those on the pivot flank slacken the pace.

When the charge sounds, the base drops into a walk till picked up by the line.

It would not do for the colonel or the field officers to lead, because their services are required elsewhere.

They must watch the movements of the enemy and look after their troops.

The majors ride on the prolongation of the line given by the base, and repeat the signals by pointing with their swords in the new direction: this is done by all the officers.

Thus without a word being spoken a line could be led at once into the direction required, and the principle could be carried out with brigades and divisions.

The line is advancing at a trot—a change of direction is made half right by the base.

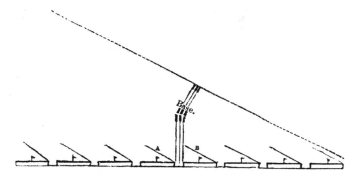

All the troops wheel into the new direction, and those on the right of the centre drop into a walk, and incline to the right.

The base troop A on the left of the centre keeps up the pace of the advance, the trot; all troops on its left increase to a gallop and drop into the trot when in the alignement.

The base troop B, and all on its right, keep on at a walk till the line comes up to each in succession, when they resume the trot.

At a walk the troops on the right would halt, those on the left come up at a trot.

In the advance in line, the *direction* of the advance is the important point, and not the *perfect alignement* of the squadrons. So long as every horse's head is turned straight towards the point of attack, it really matters little whether one squadron be or be not a few horses' lengths in advance or in rear of the line. This is the common-sense view of the matter. Keep your alignement as perfect or as neat as you can, but do not make sacrifices to obtain the end, and, above all, think of your *direction*.

10

CAVALRY TACTICS

"Good riding and skill in the use of the sword are the fundamental points of all cavalry tactics."—General De Brack

GENERAL RULES

I. Always form so as to have your front clear as well as your flanks, unless protected by woods, gardens, enclosures, etc., occupied by your troops.

II. If you have a position to defend, form in rear of it.

III. Never attack without keeping part of your strength in reserve.

IV. Let your formations extend in depth, not in length; with large bodies of cavalry, attack with one-fourth of your strength; protect your flanks, follow up successes, or retrieve reverses with the remainder.

V. Conceal your movements where the nature of the ground will admit of it; where not, still endeavour by your speed to surprise the enemy.

VI. Never put a line of cavalry threes about to retire in the face of an enemy; they get out of hand, and are not easily stopped.

VII. Never attack with more troops than necessary to the object in view.

VIII. When possible, always reconnoitre the ground over which you are going to charge, to ascertain that there are no impassable obstacles.

IX. Charges on a large scale should seldom be attempted against masses of troops of all arms, unless they have been previously shaken by the fire of artillery.

X. Always watch for and try to seize the right moment for attack.

XI. Cavalry should not be *brought into action too early* in the day.

XII. Cavalry is best employed on an enemy's flanks.

N.B. Three of these Rules were neglected by Brigadier Pope's brigade of cavalry at the battle of Chillianwalla, namely, Rules III., IV., VI.

Rule III. They had no reserve.

Rule IV. The four regiments of cavalry composing the brigade were all formed in one line.

Rule VI. The line was put threes about, and could not be stopped.

Cavalry versus Cavalry.—The most difficult position a cavalry officer can be placed in is in command of cavalry against cavalry, for the slightest fault committed may be punished on the spot, and a reverse lead to the most disastrous consequences.

When acting against the other arms, a failure is by no means irreparable, because their movements are comparatively slow.

Cavalry caught in the act of forming must generally be overthrown if its adversary profit by the opportunity.

Avoid, therefore, all manoeuvring when within reach; advance in the order of attack; and if forced to make a change in your dispositions, do so quickly, and so as to keep the front of your leading troops clear, in order that they may charge should the enemy attempt to interfere.

When precluded by the nature of the ground from forming line before advancing to the attack, advance in columns of little depth, in order to form quickly and unexpectedly on that point where you wish to strike; but to execute a movement of this sort in the face of an enemy is dangerous, and requires confidence and skill, unless supported by the fire of artillery.

A cavalry officer must know how to mask his intentions, and how to show a front quickly under all circumstances; for seldom do you come upon the enemy in the precise position you expected to find him; nor will he easily give you an opportunity of carrying out a preconcerted plan of attack;—the slightest move on his part will oblige you to make a counter one, and that without *hesitation*.

The eye must scan the *distance* between you, and you must feel certain that you have time to carry out a movement before attempting it, for to be too late would entail defeat.

And here I would observe that the eye, both of men and officers, ought to be previously and assiduously practised in measuring distances. One man well accustomed to use his eye in this way will tell at a glance the distance between himself and any given object, or will err only by a few yards or feet; another man, without the habit, will make the wildest guess. In the military schools of France they make this training and practising of the eye a regular part of the officer's education. So ought it to be with us, although, apparently, no one seems to have given the matter a thought. To the cavalry officer the eye is the most valuable of all the organs, and in him everything ought to be done to develop its faculty and power.

Marching on a parallel line to that of the enemy in open column, to wheel up and charge, is a simple good movement, and sometimes takes the enemy by surprise; but it is not always safe, for you expose your flanks. In such a case, if the enemy approaches, you must at once wheel into line those troops that are opposite the foe; the rear troops continue their march behind the line and form up in succession: or you advance to the charge with the troops wheeled up, whilst the rear of the column wheel into line also and form your support.

An officer leading cavalry into action should name some officers or non-commissioned officer (well mounted) to precede the column and reconnoitre the ground over which the attack is likely to be made.

If the enemy is in eight, these officers should gallop straight towards him, approaching as near as they can without running the risk of being captured, and make themselves masters of any feature in the ground which may be turned to advantage, or prevent their own troops from attempting to charge over obstacles which might lead them to destruction.

Cover your movements and protect your flanks with skirmishers, and reinforce them according to circumstances when within reach of the enemy. Under the protection of your skirmishers watch the adversaries' movements. The officers with the skirmishers must keep a good look-out on the flanks, so as not to overlook any movement towards them, favoured by the ground, or covered by a village or other enclosures. By a manoeuvre of this sort the French cavalry were defeated by the Austrians at Wurzburg. The Archduke Charles sent fourteen squadrons of hussars to turn a village whilst the French were advancing to attack his cuirassiers; the hussars allowed the French line to pass the village in their advance, then galloped in on their rear and did great execution.

When ordered to attack, take the initiative, and, when advancing against a superior force likely to out-flank you, keep troops in reserve behind your flanks with orders to act as circumstances require. Thus having secured your flanks, and being backed up by a reserve, fall on without hesitation at the favourable moment—such as a change in the dispositions of the enemy, when they are in unfavourable ground, or when they are suffering from the fire of your artillery.

The enemy may try to take you in flank or surround you, but whilst taking ground to the right or left for that purpose he exposes his own flank, and the troops behind your flanks must take him in the fact whilst you charge home.

"A manoeuvre," says De Brack, " which is often attended with success, when two lines are watching each other, is to make a squadron break forth in column from behind one of your flanks (or the flank squadron itself), and press forward as if to turn the enemy's flank. They will immediately wheel into column to prevent this; then, at once, sound the gallop, and bear down upon them. This manoeuvre is the art of war in miniature."

A fine example of steadiness and resolution was shown at the battle of Blenheim (1704) by a British regiment of cavalry (the carbineers) in the early part of the action, just after the great attack on Blenheim village, which, though our infantry could not force it, was eventually cut off as it were from the French line by the occupation of the line outside, and a large body of French infantry was pent up within its enclosure, and rendered useless.

"The Duke," says Kane, "having thus secured himself on that side, ordered Colonel Palmes with three squadrons to pass over the brook, who, meeting no opposition, drew up at some distance from the marshy ground, and gave room for our lines to form behind him. The Duke followed Palmes: the mills were attacked, but those that were in them set them on fire and made off. Both the cavalry and infantry which the Duke kept with him (not above ten squadrons and twelve battalions) passed over as well as they could, and formed as fast as they got over. Tallard all this while, as a man infatuated, stood gazing, without suffering either great or small shot to be fired at them; only when he saw Palmes advanced towards him, he ordered five (some say seven) squadrons to march down, and cut those three squadrons to pieces, and to return. The officer that commanded the French squadrons, as soon as he got clear of the line, ordered the squadrons on his right and left to edge outward, and then to wheel in on the flanks of Palmes; which Palmes perceiving, ordered Major Oldfield, who commanded his right squadron, and Major Creed, who commanded that on the left, to wheel outwards, and charge the squadrons coming down upon them; and, not in the least doubting their heating them, ordered them, when they had done that, to wheel in upon the flanks of the others, while he at the same time would charge them in front. Accordingly everything succeeded, though with the loss of some brave men, and Major Creed, among them; so that these three squadrons drove their five, or seven, back to their army. This was the first action in the field, which took up some time, and gave the Duke an opportunity to form his lines."*

With a large body of cavalry it is difficult to gain an enemy's flank, but with small detachments the opportunity often occurs. Cavalry officers, when they see a good chance given them by the enemy, should never wait for a better one.

If the enemy's cavalry have to clear a ditch, a hollow way, or any other obstacle in their front, let them attempt it, and fall on before they have recovered their order or resumed the speed of then advance.

A cavalry engagement is seldom decided by a single charge, but the advantage remains with those who have the last reserve of fresh troops at their disposal: this generally turns the fortune of the day.

Large corps of cavalry are now generally formed in three lines—the first line, the second line, and the reserve.

Now suppose a second line of cavalry deployed according to rule, with intervals of twelve paces between squadrons. The first line, of one

*Lord de Ros, *The Young Officer's Companion.*

thousand or twelve hundred horsemen, with horse artillery, are defeated and pursued by the enemy. Shouting at the hotly pursued fugitives will never make them diverge from the straight line, their shortest and safest road, nor induce them to ride round the flanks, nor can they by any possibility escape with guns and tumbrils through the small openings left in the second line.

The second line could not advance unless its front was clear, and, seeing the enemy rushing upon them without the possibility of charging against him, they would probably turn and gallop for it.

The first line alone should be deployed.

The second line formed in double open columns, the two flank columns outflanking the first line.

Each column could easily clear its own front and break through the crowd. Should the enemy continue the pursuit of the first line, by detaching the rear troops outwards they would fall upon the enemy's rear, and, galloping with them, destroy them in the pursuit

These columns could act at once when they had the chance, for in the confusion it would be utterly hopeless to expect one effort of the whole line at once.

Reserves in columns in rear of the centre would be in the best position to detach reinforcements when they were needed.

If the first line succeeds in the attack, the columns of the second are ready to advance and support, whereas if in line they must first break into column, and thus lose much precious time.

When a first line is not considered strong enough to attack the enemy, the reserve joins the second line: both deploy and advance together, the first line passing through their intervals to the rear, and forming the reserve.

The difficulty of advancing with a long line, and the danger of being driven back at all points at once, make it advisable to use échellons instead of attempting to charge on a large front in line.

Échellons are useful when debouching from a defile to support troops already engaged or defeated; also to pass through intervals and attack a line made unsteady by the fire of infantry or artillery, and fall on without giving them time to restore order.

The échellon should not be formed of less than a wing, or it would hardly take effect against an enemy's line: further, wings can break into échellon, columns of troops, contiguous double and single open columns, and thus become handy in every way.

Some of the advantages of an advance in échellon are,—

You at once avail yourself of an opening given by the enemy, and charge without waiting to form line. You can attack one of his wings, or any other point in his line, without engaging all your troops at once; the enemy cannot attempt to attack the flank of the charging échellon without exposing his own to your succeeding échellon.

If the first one is defeated, you support it; whereas, if successful, you are almost sure to defeat your adversary, because, when the remainder of his line advances against you, the first victorious échellon falls upon his flank and rear.

After a successful charge, rally forward, picking up the stragglers as you go on. If you sound the recall you lose time and bring the men twice over the same ground,

When rallied, attack the flank which has been exposed by the defeat of your opponents.

General Boussard at the battle of Llerida charged the Spanish horse, formed on the left of the line, with the 13th regiment of cuirassiers, and overthrew them; then bringing up his right shoulders, he fell without hesitation on the Spanish infantry, which, taken in flank, was ridden over, and 5600 of them were made prisoners.

Cavalry seldom meet each other in a charge executed at speed; the one party generally turns before joining issue with the enemy, and this often happens when their line is still unbroken and no obstacles of any sort intervene.

The fact is, every cavalry soldier approaching another at speed must feel that if they come in contact at that pace they both go down, and probably break every limb in their bodies.

To strike down his adversary, the dragoon must close, and the chances are he receives a blow in return for the one he deals out.

There is a natural repugnance to close in deadly strife. How seldom have infantry ever crossed bayonets! some authors say *never*! and cavalry soldiers, unless they feel confident in their riding, can trust to their horse, and know that their weapons are formidable, will not readily plunge into the midst of the enemy's ranks.

Lines advancing to meet each other have shown hesitation at the same moment, thus:—

In the retreat of our army from Burgos, three squadrons of French chasseurs charged some squadrons of our rear-guard; these advanced to meet them; both lines pulled up close to each other and stood fast, till one of the French made a cut at the man opposite to him, upon which both lines instantly plunged forward and engaged; the colonel of the chasseurs was killed, most of his officers were wounded, and the French were driven back with a heavy loss.

The flanks of each troop should be closed up by officers, for it is generally from the flanks that men ride to the rear.

Cavalry with an Army.—When cavalry is assembled in large bodies, it is for the purpose of striking a decisive blow.

Natural obstacles, as well as the difficulty of breaking through masses of infantry, often prevent cavalry from closing with the foe.

Infantry now fight on all sorts of ground; a level plain is no longer sought for as a field of battle, and in an enclosed country cavalry is required to act as an auxiliary to infantry, or a protection to artillery, and must be scattered in small bodies to effect this.

Thus, to be prepared for all emergencies, it has been customary to add a regiment of cavalry to each division of infantry, to serve as a support in battle and to seize upon and follow up any advantage gained.

The remainder of the cavalry is then formed in reserve, either in rear of the centre or both flanks of the second line.

Cavalry must act independently once it starts for the charge: the fire of infantry and artillery may pave the way before, or even during the attack, but there must be no hesitation when the cavalry advances to the attack: kept back till the right moment, when it breaks forth, let it be *to fight*.

The battle of Möckern, fought on the 16th of October, 1813, affords a striking example of the results obtained by a small body of horsemen well commanded and brought forward at the proper moment.

This battle, so glorious to General Yorck's corps, was more particularly so to Colonel Sohr, who, whilst the battle raged around him, waited steadily till the enemy gave him a favourable opportunity, and, charging that instant with determined bravery, turned the scale in favour of his country, and chained victory to the Prussian standard.

The possession of the village of Möckern was desperately contested on the part of the French, and, though the Prussians at last carried it, they could not pass out on the other side, the enemy having established powerful batteries on the heights beyond.

One battery of fifteen heavy guns spread death and destruction in the village itself; fresh troops pressed forward on both sides to the attack, until on the Prussian side one brigade only was available (it was commanded by Colonel Steinmetz). These men General Yorck threw into the scale: they so far re-established the fight as to enable the Prussians to hold their own in the village and maintain the ground on the right, but they gained no decided advantage.

What was now to be done? The only fresh troops on the ground were the cavalry reserves under Colonel Jürgas, in position near Wahren, but it was no use ordering them up whilst the enemy was fresh and unshaken.

Sohr with three squadrons of hussars had been detached early in the day to support the right flank of the infantry, and stood in column between two ridges to avoid the fire of the enemy's artillery; their shells, however, obliged him to shift his ground a little to the left of the road.

At this time the French again drove part of the Prussian infantry back into the village, and established themselves in force in the houses and enclosures.

General Yorck rode up to Sohr and said, "If the cavalry cannot do something now, the day is lost! advance, Sir, and charge." Sohr drew his attention to the fact that he had no reserve to fall back upon in case of defeat, the cavalry being still in rear of the centre." An aide-de-camp was sent off to order them up, and the General, as he was riding off, directed Sohr to try and make our infantry hold their ground till the cavalry came up; whilst doing so. Major Schack came again with an order to charge. Sobr answered, 'Tell the General I am only waiting a favourable opportunity; this he must allow me to choose; I will then charge, and on the honour of a soldier I'll charge home.'"

The enemy had now advanced so close that a charge of bayonets against the Prussian infantry was expected every moment. Wrapped in clouds of smoke they came on firing, and Sohr could tell only by the whistling of the bullets that they were quite near enough. He passed through the line of retreating infantry, formed, and with three cheers burst upon the advancing enemy, rode over and dispersed them, pursuing them into their own batteries. He captured six pieces of cannon at the first onset.

The enemy's cavalry, indeed, rushed down to the rescue, but a regiment of lancers from the reserve had now joined, and they advanced together and overthrew the French, the lancers pursuing them, whilst the hussars attacked the infantry formed in squares to receive them, broke into three squares in succession, and captured nine pieces of cannon; after which they joined in the pursuit and followed the enemy close up to Leipzig, inflicting on him a severe loss in killed, wounded, and prisoners.

An officer in command of cavalry must know what he intends doing; this he must carry out with energy and resolution, for under vacillating commanders no cavalry, however brave, will do any good in the field.

A commanding officer should not remain in the immediate neighbourhood of his corps, but ride forward to watch the progress of events, and be ready to seize a favourable opportunity for employing cavalry, as well as to guard against surprise; and in either case, unless he have orders to the contrary, he should act at once.

The reserves should be placed out of sight and sheltered from the fire of the enemy, ready to follow up a first success; without them no pursuit can be kept up, for, when exhausted, a handful of fresh troops would destroy the pursuers and turn victory into defeat.

As a general rule, cavalry should not be brought into action too early in the day, unless, indeed, a favourable opportunity offers; then, of course, make the most of it. It should be held in hand to decide the victory, to retrieve lost ground, to cover a retreat and save the army from the loss of artillery, etc.; and for any serious exertion fresh men and horses are absolutely necessary.

Attack and defence of positions.—Before attacking an enemy's position, as much artillery as can be collected should concentrate its fire on that part of the enemy's line on which the charge is to be executed, and the guns placed in such a manner as to allow of their keeping up their fire during the cavalry advance. If the charge fail, they draw off to the flank, and the artillery re-opens its fire.

To repulse an attack of this sort, form part of the cavalry ready to attack them in front, but so as to avoid crossing the line of fire of your own guns; with the remainder of the cavalry take up a position to attack them in flank. Wait till the enemy has suffered from the fire of your infantry and artillery in his advance, then fall upon him in front and flank simultaneously. If you succeed in overthrowing him, keep him going, give him no time to reform, drive him pell-mell into his own position, enter it with him, and spread confusion everywhere; your second line following up the success, and detaching parties to attack the gunners wherever they see an enemy's battery, whilst the reserves follow slowly and prepare to cover your retreat in case of accidents.

Throwing cavalry forward on an enemy's front succeeds when the enemy is disheartened by previous defeats or disorganized through other causes. But to throw cavalry forward against an enemy's position is a dangerous experiment, particularly when they have cavalry in hand to let loose upon you. Should the attack fail, it may then result in the total defeat and destruction of the attacking force. Sometimes it may be necessary to make the attempt. For instance, at the battle of Eylau, Augereau's corps, formed in columns of attack and supported by a powerful artillery, advanced against the centre of the Russian position. Augereau, both his generals of division, and several other general officers, were wounded, some of the regiments nearly annihilated by the destructive fire of the Russians and their charges of cavalry and infantry. The entire loss of this corps (the 7th) was imminent, and its total destruction was averted only by the timely advance of the French cavalry. Napoleon sent them all forward, under Murat and Bessières, to disengage Augereau, and make a general attack on the Russian position. This remarkable charge was executed by seventy squadrons of cavalry, the flower of the French army. They fell straight upon the Russians, broke through their two first lines: but the Russian cavalry reserves, consisting of a hundred squadrons, fell upon the scattered squadrons of the French, and drove them back with immense loss, whilst the broken lines of the Russians at once re-formed.

At Waterloo every attack of the French cavalry on our position was repulsed. These attacks were preceded by a terrific fire of artillery; they were continued and kept up with the greatest obstinacy and perseverance, yet they never gained even a partial success during the battle, and the losses they occasioned to the French were very great.

When you are the strongest in cavalry, it is then best employed on an enemy's flanks, otherwise keep it in hand under the protection of your infantry and guns, to be used according as opportunity offers.

When cavalry and guns can be spared, they may be sent round the enemy's position to fall upon his rear, but only when their march can be concealed from the enemy (unless intended as a false attack, merely to draw off his reserves in that direction); otherwise it is better to bring them into play at once, for should they go around, when they do reach the rear of the enemy he will have made his dispositions to receive them, and you will have gained nothing and lost time.

If they succeed in surprising the enemy, his cavalry should be attacked on the spot; but if his infantry be prepared, then use the horse-artillery first, to ensure the result of the charge.

Cavalry in reserve behind the crest of a hill, up which the enemy is pressing forward, should, when possible, form line before coming in sight of the enemy, and thus be ready to fall on before he can prepare to receive them.

In the pursuit of a defeated army, cavalry act in bodies, directing their efforts against such of the enemy as still keep together. With small detachments prepared to receive them in positions difficult of approach, no time should be lost; these are sure to fall a prey to the infantry. The cavalry must pass on to either side, cutting in upon the line of retreat and preventing order from being restored in the broken masses of the enemy.

In 1796 Napoleon, at the battle of Roveredo, by a well-concerted attack threw the Austrian army in disorder into the defile of Calliano and routed them. One of his aides-de-camp, Lamarois, dashed through the mass of fugitives with fifty dragoons, and, heading them, checked the retreat of the column, and thus enabled the French cavalry to secure several thousand prisoners.

Artillery is the great support and rallying-point of a defeated army; against this arm in particular the most strenuous efforts of the cavalry should be directed. Once an army has lost its artillery it cannot long keep the field.

Horse-artillery in such a pursuit not only assists the cavalry in its charges, but brings the infantry and artillery of the enemy to a stand, and thus they fall into the hands of the troops coming up.

Horse-Artillery with Cavalry is not only a powerful auxiliary in the attack, but a friend in need in the hour of danger and defeat. It can keep pace with the advance, and by pouring a destructive fire on the enemy pave the way for the cavalry to victory.

It dislodges the enemy from positions in which the sword cannot reach him, or it does fearful execution where the enemy's infantry, concentrated in masses, bids defiance to the horseman.

It checks the pursuit and gives time for the cavalry to reform under its fire.

It is of great use in passing defiles in the face of the enemy, in defending broken ground against odds; and in many other ways it is the cavalry's best assistant.

Horse-artillery should place itself so as not to interfere with the intended formations of cavalry, and always be at hand to protect them when forming.

It is advisable that the artillery should be on both flanks, or masked by the flank troops during an advance, or kept together on one flank, particularly if the column in forming line throws forward the opposite flank or forms in succession; the whole of the guns are then brought into play on the enemy till the line is ready to charge.

When there is but one troop of horse-artillery it is best to keep it together; so small a number of guns should not be subdivided without an absolute necessity.

Horse-artillery can always take up positions to engage the enemy at two or three hundred yards in front or on the flank of cavalry and be pretty safe, but it should not venture further without an escort.

The artillery officer should, as far as possible, choose his own positions and open fire when he thinks proper. Non-professional officers cannot be such good judges of range, etc., and should not interfere; thus much ammunition will be saved to the army.

If cavalry is defeated, its horse-artillery goes to the rear as quickly as possible; for it cannot at first, whilst friend and foe are going *pell-mell* to the rear, do anything towards restoring the fight; but as soon as it finds a favourable position it should unlimber and check the foe.

When the enemy's cavalry is drawn up in battle array to await the onset, and its horse-artillery is posted 200 or 300 yards in front, the guns of the attacking force advance to within 800 or 1000 yards and fire with round shot upon the artillery until their cavalry moves forward. The moment these latter come within range of grape, they direct the fire upon them; the rule for artillery being to deal with the most dangerous foe for the time being.

In a retreat it is usual to give the officer of artillery notice: he then chooses his own positions and avails himself of every advantage the ground may offer. The cavalry holds the enemy in check till the guns are in position, then falls back under their protection and re-forms to cover the retreat of the guns to the next position, and so on. When the cavalry is in motion the artillery should be in position; when the horse-artillery is in position the cavalry should be formed to repel an attack.

The retiring troops, if in good order, cannot easily be followed closely by their pursuers, who must always advance with certain necessary precautions, which take up time, and yet dare not, cannot be neglected. Those retiring occupy every favourable position to stop the advance of the enemy, and this they can often effect, without firing a shot, by destroying bridges, roads, etc.

In a pursuit horse-artillery follows the cavalry steadily at first, so as not to over-exert the horses, for while the cavalry are mixed up with the fugitives the guns are of no use; but as soon as the enemy begin to gather together for defence, or attempt to hold a position with fresh troops, the guns come to the front to start them, and slip the cavalry, who will keep them going.

Escort of artillery is not absolutely necessary when the guns are acting with a column of cavalry, unless they move off to a flank or beyond 300 yards from the column; then two troops should accompany the battery, whether consisting of one or two troops of horse-artillery.

The escort protects the flanks and rear of the artillery; the guns protect their own front.

It should not form directly in rear of the guns, but on the right or left rear, or on either flank; always the outward flank when the guns have troops on the inner one, and (when feasible) covered from the fire and view of the enemy. When the escort is necessarily exposed, the commander should watch the fire of the enemy's guns, and change his position from time to time to avoid being made a target of.

When the escort is not sufficient to repel a sudden attack on the guns, the officer commanding the troops or squadron nearest to it should at once fly to the rescue without waiting for orders; and this principle of mutual support should be encouraged under all circumstances between cavalry and horse-artillery.

These remarks on horse-artillery with cavalry merely intended to give a general idea of the way in which these two arms support each other in war, and to stimulate their spirit of enterprise by reminding them that, if they combine the use of these two arms skillfully, they are certain of obtaining distinction in the field.

The attack in skirmishing order should be practised because it is useful in various ways.

If, when advancing towards an enemy, he turns to make off, attack in skirmishing order with part of your troops, because the horses have more freedom of action when not in the ranks; each man, acting for himself, makes for his nearest foe, and is more likely to catch the enemy and inflict loss upon him than if pursuing in a body.

Artillery may be attacked advantageously in skirmishing order, and with very few men. They advance straight towards the guns till within range of grape, then put the steam on, gallop as quick as they can, opening from the centre, and making for the flanks of the battery.

The gunners are thus obliged to alter their aim and lose time. If there is much dust, or the smoke has not cleared away from the front of the battery, they do not observe the change of direction, continue firing straight to the front whilst the horsemen are pouring in upon the flanks, and immediately attack the gunners to prevent them from serving the guns, whilst the cavalry reserves advance, part of them charging the escort, the remainder securing the battery.

The most advantageous moment to attack is when the arms are unlimbering or limbering up: a few moments gained being of the greatest importance to the attacking party, who must always dash in at their best speed as soon as the signal is given to charge.

If, after getting into the battery, they see a superior force of the enemy coming to the rescue, they should endeavour to do as much damage to the battery as possible by cutting down the drivers, severing the traces, etc., and then trust to the speed of their horses for safety.

Reconnoitring.—This is one of the most difficult problems in war, to know how to gain reliable and sufficient information before it becomes necessary to engage an enemy. I shall therefore endeavour to give an outline of the duty of cavalry when employed for this purpose, but to enter into details or go deeply into this subject would carry me quite beyond the limits of this work.

General Rules

1. Press forward quickly, and avoid engaging the enemy when it may retard your advance.

2. Drive in the enemy's picquets, and press on to the point ordered, retiring as quickly as you came on.

3. Unless a favourable opportunity offers of surprising a party of the enemy, avoid all conflicts, for it is not the object in view.

4. Where the enemy stand in your way, no danger should turn you from the point aimed at.

5. Supports should always be moved up in the line of retreat, and bridges or defiles occupied by infantry for the reconnoitring party to fall back upon.

It is often necessary to send out powerful detachments, in order that the officer commanding them may push forward far enough, or engage the enemy and force a reconnaissance, and also to enable him to detach patrols and exploring parties in different directions, without being so much weakened by these necessary measures as to run the risk of being attacked and driven in without attaining his object. To determine the amount of force which ought to be detached on a reconnaissance is always a nice point.

A large army (say of 60,000 or 70,000 men), marching in column near an enemy, is obliged, when short of cavalry, to make one corps cover the flank of another, a third one cover the flank of the second, and so on, till the army is often scattered into detachments, and spread over so much ground, that a bold move on the part of the foe is not seldom successful in beating them in detail.

Troops are more easily detached than assembled: when together they have little to fear from any manoeuvres: therefore, when the nature of the country admits of it, two or three thousand horsemen are detached to the front and flanks to feel for the enemy, whilst the army continues its advance in concentrated strength.

Cavalry thus employed must go well ahead of all the columns; it must not hang together, but spread for miles in all directions; so that nothing shall escape detection.

It must penetrate everywhere and look up the enemy at all points, never forgetting that to fight is the means, not the object; but when other means fail, then no danger should daunt them from discovering what the enemy attempts to mask from their sight.

For instance, the enemy is in force behind a ridge, and occupies the side towards you with skirmishers to prevent your discovering what troops he has. These skirmishers must be driven in, or, whilst they are attacked, a few well-mounted men or officers should dash through, ride over the ridge, and reconnoiter the enemy's position, trusting to the speed of their horses to regain their party. For a dragoon with a good horse under him should fear no odds in an open country.

11

THE MARCH—THE CAMP—THE BIVOUAC— THE OUTPOST—THE SKIRMISH

THE MARCH

In the campaigns of the last great war in Europe it was no uncommon occurrence to see cavalry arrive on the field quite crippled, having lost half their numbers before a shot had been fired, the remaining horses being in such wretched condition as to be totally unfit for active service.

This arose partly from want of sufficient and good forage, as also from the fact that they generally began their marches without any previous training; the horses, distressed at the start, never recovered themselves, and died off by thousands.

It is well known that neither horse nor man can undergo fatigue suddenly without suffering from the effects of it. Both should be brought gradually into working condition.

Collecting large bodies of cavalry together, long before they are required to act on the battle-field, is a great mistake.

Such corps have to go greater distances in dispersing to their billets or camps, and the same in reassembling in the morning. There is great difficulty in procuring forage for so many. Sickness is caused amongst the horses from the mere fact of great numbers being collected together. Infection spreads far and wide, and the difficulties of marching steadily are greatly increased in consequence of the frequent checks and constant closing up of the columns of route. All this must largely contribute to render cavalry inefficient for the field.

I am convinced that by observing a few very simple rules the greater part of these serious losses might be prevented, and horses and men be brought into the field of action in excellent health and condition.

Having previously got the horses into good working order by degrees, or by making the march of one day a little longer than that of the preceding day, then begin your real march, moving in small bodies, with an interval of one day's march between each body. If fresh from cantonments, begin the marches very quietly, increasing the distances by degrees; thus having got your horses into hard-working condition, when you approach the enemy they can be easily drawn together by forced marches; and if the army win the day, the cavalry will be in such order that the enemy will with difficulty escape from their pursuit.

Until necessity requires it, the cavalry corps should be detached; the horses will thrive, they will be well fed, well looked after, and will not be found wanting in the hour of need.

With bodies of cavalry on a march, the leader of each detachment having an open interval (whether squadrons or troops) should lead at steady pace, not increasing and slackening as the rear files of the squadron in front of him do; for in that case the unsteadiness communicates itself to the rear, whereas it should stop at the end of each squadron.

The pace a slow trot, about six or seven miles an hour, the men rising in their stirrups, and walking the horses up and down hill. The horses get in early to their food, they are groomed and better looked after, and have more time to refresh.

If you walk all the way, the horses are kept saddled many hours more than is necessary; the men get tired, sit unsteady in their saddles, and the horses get sore backs.

The crawling kind of march really fatigues men and horses much more than a march at a smart trot. Let any man ride a journey of twelve or fifteen miles at a walk, without ever breaking into trot or canter, and tell us when he dismounts how he feels. The horse is always distressed by being too long under the saddle, even though he stand stock still all the while.

Cavalry is often obliged to march slow in order to keep company with other troops; but otherwise it should go on steadily at a quiet trot, and keep the horses as short a time as possible under the saddle.

Before starting on a march, the shoeing of the horses must be looked to, and at all times, whatever the object of the march, the proper military precautions must be attended to.

The order of march should be such as to be easily transformed into the order of battle.

Advanced Guards.—All troops marching are preceded by advanced guards, to cover the front of the column, give timely notice of danger, and prevent surprise.

The strength of an advanced guard is in proportion to that of the column it precedes; and as a large body of troops requires more time to make arrangements for receiving an enemy than a small one, the advanced guard should precede the column it covers at such a distance and in such a strength as in case of attack to give time for the main body to form before the enemy can be upon them.

In time of war always procure a guide (by force if necessary).

Patrols should constantly push forward to front and flanks to feel for the enemy, search the side-roads, farm-houses, woods, etc. etc., and look into any place likely to conceal troops. Always secure your flanks before entering defiles, by taking possession of the heights, or other commanding positions, near them.

When one army is in march to attack another, the advanced guards are generally used to obtain a knowledge of the enemy's position; this they do by driving back the enemy's advanced posts the moment they fall in with them, following them up quickly, and reconnoitring the enemy's position and strength from the ground his picquets have abandoned.

The Rear Guard protects the rear, makes stragglers keep up, and prevents them from plundering.

The place of the officer commanding the rear-guard is with the last detachment of it.

During a retreat the duties of a rear-guard are most important, and to prevent its becoming a rout the troops must be steady and well commanded.

They should halt in all defensible positions, but never at the entrance of a defile, and always on the further side, taking possession of the outlet to check the enemy.

The Camp

Officers are often required to mark out camps for troops in the field (*Figure 2*). I will endeavour to convey to them an idea of how they ought to act when employed in this service; and as no camps can be traced until the outposts are taken up, it will be well to give a sketch of that duty at the same time. To enter deeply into it would be superfluous, for we have many good books on the subject—as *Manual of Outpost Duties*, by Lord Frederick Fitzclarence; *On Outposts and Patrols*, by Major-General L. B. Lovell, K.H.; *On Outposts and Patrols*, by Lieut.-Colonel Von Arentschildt; *Outpost Duties*, abridged from the German, by Lieut.-Colonel Short, etc. etc.; and many others.

General Rules for Camping

1. The object is to put many men in a small compass.
2. To give them rest and as much comfort as possible.
3. To place them in such a manner that they can get under arms without loss of time.

To attain this the tents or huts must be raised systematically and according to rule. Thus no room is lost, the men can attend to their duties without confusion, and easily find their way about the camp.

Camps ought to be on the highest, or in front of the highest ground, with plenty of room, an open view, with the flanks protected, with easy communication from flank to flank and front to rear, and with wood and water in the neighbourhood.

When near an enemy, you should as far as possible encamp in order of battle, or in such a manner that you can easily get your troops into position. Where circumstances do not require this precaution to be taken, you then encamp with a view to the comfort of the men, and in such way as to give them as much rest as possible.

Camps are therefore of two sorts—Marching, and Fighting, Camps—according to the presence or absence of an enemy.

Tents are seldom used in Europe, except for forming camps of instruction: they have long since been discontinued in time of war, because they retard the movements of an army in the field; and also because, in a highly cultivated country, shelter can generally be found,

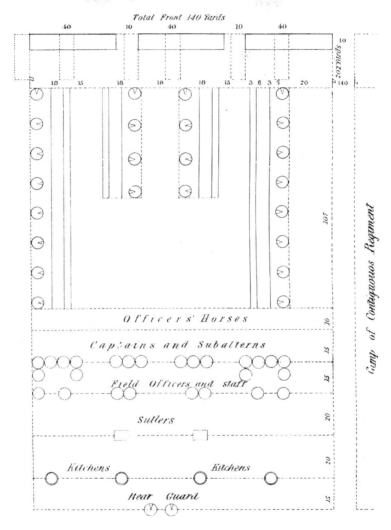

FIGURE 2. A CAMP MARKED OUT.

or material procured with which to construct huts for the men: these are much the best description of cover.

Marshal Oudinot introduced a species of tents amongst the French troops, which are set up on the muskets and bayonets: these can be raised and struck in a moment, but give little shelter against wind and weather, and add much to the weight which the soldier has to carry.

Bivouacs have the advantage over other plans of camping—that the troops can be at once got under arms, and, therefore, all outposts bivouac; but the army should not do so except in case of necessity, as the exposure to the night-air, the cold, and damp, destroys the health of the men.

In a camp the length of front is calculated according to the number of regiments or squadrons, adding the intervals.

In hilly countries the valleys are often left as intervals between the camps. This does not affect their safety, as each camp is strong in its own position, and can protect each other from attack.

Cavalry or infantry encamp either in line or in open column. The infantry pile arms in their front.

Cavalry should never camp in the front line, but either in rear, or in rear and on the flanks of the infantry; so that, in case of attack, it has time to turn out and move up to the support.

The artillery encamps with the brigades to which it is attached; the artillery of the army in rear of the centre of the second line. Precautions must be strictly enforced against fire.

Artillerymen encamp fifty yards in rear of their batteries, and place sentries to enforce all necessary precautions for the safety of the ammunition-waggons, etc. etc.

The artillery of reserve, the parks, and ammunition, are placed in rear of the army, formed in one or two lines, and surrounded by pickets and sentries; the artillerymen and horses in rear or on the flanks of the line, at one hundred and fifty or two hundred yards' distance.

The forges at such a distance as to avoid all danger from the sparks. All camps have outposts with outlying and inlying pickets. The main body alone are allowed to rest undisturbed. Further measures must be taken to provide wholesome food and water, and to keep the camp clean, and free from beggars, vagrants, improper women, etc.

The Camp in Time of Peace

When the only object is to concentrate large bodies of troops for exercise, you should pick out a site where all the necessaries can be procured close at hand, in order to save the men from unnecessary fatigue.

Avoid wet, sandy, or swampy ground. Pitch the camp on a slope to carry off the rain and all impurities. Place it near villages, woods, rivers or streams, from which you can obtain wood, water, and straw.

No washing, bathing, etc., should be allowed in streams near the camp, except below it, so as to keep the water pure for use. In these sorts of camps the men are generally under canvas.

Marching Camps

The same rules apply as for camps of exercise except that, as the troops pass the night only in them, they generally bivouac or shelter themselves from the weather in the best way they can, with branches of trees or thatches of wicker-work, straw, etc.

The officers who mark out these camps endeavour to have the forage, wood, and straw brought in, and thus save the men the necessity of going out for them after a long march. They can then cook at once and make arrangements for the night.

Fighting Camps

Are generally placed exactly in the contrary positions to those already described; for instance, they should be on high ground, whereas the others follow the course of the streams into the valleys, which are naturally more thickly inhabited, better cultivated, and afford the necessary supplies.

Camps in the field depend upon the line of operations and intentions of the commander; all other considerations are sacrificed to the higher objects in view.

They must be placed in advantageous military positions. To choose these, officers of experience are sent forward with reconnoitring parties; they are guided by the following rules:—

1st. That the flanks are protected by mountains, ravines, water, or villages, which oblige the enemy to go far to turn them.

2nd. That the front of the camp is clear, as far as guns can range.

3rd. That the camp is not intersected by water, hollow ways, woods, etc., or anything which would give the enemy opportunity or cover for an attack, or the means of passing unobserved to the rear.

4th. That according as the army is formed in two or more lines, the ground shall afford sufficient depth for them to camp with the proper distance between each; and give room for the parks and ammunition in rear of all.

5th. That the ground be dry and have water and wood at hand.

Though often necessary, it is not always possible to entrench a camp so as to secure it from the enemy, because the time is generally too short to cover the approaches with regular works. In such cases the roads and the approaches are barricaded or made difficult of access, and these precautions should never be neglected; they enable you to hold the enemy at bay till the troops get under arms. In order to have time to take these necessary precautions for the safety of the camp, it should, when possible, be occupied early in the day; this rule ought not to be departed from, except after an engagement or under other peculiar and unavoidable circumstances.

Man and beast suffer if a camp is occupied at night No one knows his whereabouts; the men are worn out with fatigue, going in search of food, wood, water, and only begin to cook when they should have been at rest, to enable them to meet the fatigues of the morrow. Horses, in the same way, are not fed till late in the night, from which they suffer more than from long marches.

To Mark Out a Camp

Two non-commissioned officers, with one private per troop, are sent under an officer. He is furnished with a return showing the strength of the regiment or regiments, and receives instructions from the quarter-master-general's office as to where the corps will encamp.

The usual way of encamping is in open column of troops, the tents or huts in lines perpendicular to the front of the camp.

Cavalry camping in line requires double the extent of its front, in consequence of intervals between the horses to enable the men to groom and handle them; it is therefore only resorted to when the ground will not admit of any other plan.

In marking out the ground for a regiment to camp on in column, the front line of the camp is the exact extent of the regiment in line, with intervals.

The depth of the troop lines is calculated according to the size of the tents or huts and the strength of the troops.

The officer must see all the lines paced off and marked with flags.

The front computed for a mounted troop-horse in the ranks is one yard, and when picketed four feet. In pacing off the ground for the troops the officer will therefore calculate the extent of front as follows:—

For a regiment of four squadrons, each of 160 horses, extent of front of each squadron 80 yards. Add to this three squadron intervals, each computed at one-fourth the front of the squadron, namely, 20 yards.

Total front 80 x 4 = 320

Intervals 20 x 3 = 60

Total 380 yards, the extent of the front line (*Figure 3*).

When troops are weak in numbers, the best mode of encamping will be in open column of squadrons.

It is his duty then to meet the regiment and lead them into camp, forming them on the front line; it is then pointed out to the men that on this line they form in case of alarm; they are then wheeled into column of squadrons, the ranks opened and filed from the inward flanks to their places in camp.

The Bivouac

When in bivouac, as soon as they have dismounted and fastened their horses, each man fixes his sword in the ground, in rear of his horse, placing the carbine against it; he then takes off his belts, etc., and hangs them on the sword-hilt, and behind them again he places his kit and sleeps there at night; otherwise, if required to turn out in a dark night, he could not find horse or arms.

The whole subject of the bivouac is one of vital importance. Some officers have been known to ruin half the horses of a regiment in one

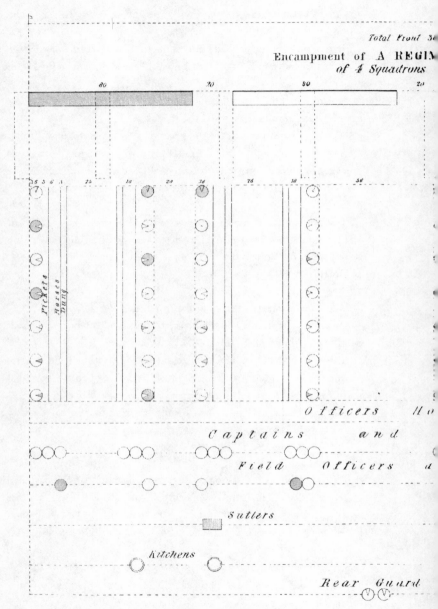

FIGURE 3. ENCAMPMENT FOR A REGIMENT OF FOUR SQUADRONS.

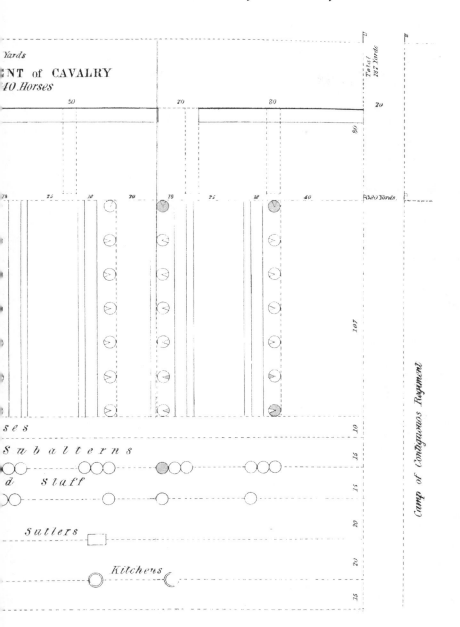

position of this sort because they knew not how to choose their ground. De Brack says,—"Of two leaders of pretty equal merit, one of whom is a good and the other an indifferent selector of bivouacs, the first, at the end of a campaign, will be able to show a numerous and well-mounted body, while the other will have only a few half-starved horses."

All huts ought to be constructed with the door or opening towards the horses, in order that the men may always have their eyes upon them. Let it be always remembered that the fire-arms must be removed from the horses the moment you dismount, for otherwise, should your horses take to rolling, your fire-arms will be broken or lost.

Outpost Duty*

Outposts have a double object: to watch over the safety of the army, and to observe the movements of the enemy.

When a corps forms the advanced guard of an army, then the chain of outposts is formed in front of that corps; but where there is no advance-guard the chain of posts is established immediately in front of the position which the army has taken up,

When the outposts are at a distance, and positions lie between them and the army which would he useful for them to fall back upon, or which would assist the army in maintaining its ground, such places are occupied by detachments from the army, and not from the troops on outpost duty.

The necessity of outposts to all troops in the field, whether it is a small detachment or an army, must be equally obvious. Men cannot stand ready under arms day and night to resist an attack; the wants of men and horses must be satisfied; they must have rest, or they cannot fulfil their duties: thus every position, of whatever kind it may be, is surrounded by a chain of guards to protect it from surprise, and to give rest and security to the occupants.

---

*Lord Frederick Fitzclarence, in his excellent *Manual of Outpost Duties*, gives a private journal of General Crawfurd's admirable outpost operations on the Coa and Agueda in 1810, by Major-General Shaw Kennedy. This journal teaching by example, and abounding in matter suggestive of reflection and comparison, ought to be diligently studied by every young officer. But the whole of Lord Frederick's concise, compact manual is of value to every man who would be really a soldier.

Outposts must be so placed that every movement, more particularly an advance on the part of the enemy, shall be at once detected; that nothing can pass unobserved between them into their own camp; and that they can hold their ground sufficiently long against an enemy to prevent their troops from being taken by surprise.

Thus troops on outpost duty are disposed of according to the different parts they have to act—namely, those who watch the enemy, and those who in case of an attack endeavour to stop him.

The first, called videttes, are pushed forward to a post whence they can overlook all roads leading from the enemy's side; they are supported by a non-commissioned officer's post, and by the outlying picquet, who are supported in their turn by the reserves.

Time is the great object to gain when outposts are attacked, consequently the reserves should be sufficiently strong to hold the enemy in check and cover the approaches to camp.

Troops for outpost duty are selected according to the nature of the ground: with an army they are generally composed of light troops of all arms— namely, horse artillery, light cavalry, and light infantry. In an intersected country the infantry is chiefly used; in an open country the cavalry. In both, the cavalry furnishes patrols to the front and flanks, and generally occupies the high roads. The cavalry pushes videttes forward beyond the infantry chain, for these can gallop back, whereas infantry soldiers might be cut off and made prisoners.

The reserves should be composed of infantry, cavalry, and artillery; how far the chain of outposts should extend depends upon the strength of the army, the nature of the country, and other circumstances.

But it is so easy to go round or turn detachments composed of regiments, brigades, or even divisions, and to attack them unexpectedly in rear, that in many cases it is necessary to extend the chain of out-posts all round the camp.

In countries where the people were unfriendly it has often happened that whole armies have thus been attacked in rear.

When your rear is protected by natural obstacles, then the chain of outposts should extend well to both flanks: when the flanks are open, then you must protect them by strong picquets, and send out parties of cavalry to scour the country on both sides.

The distance of the outposts is also entirely dependent on the strength of the main body and the peculiarities of the ground. The weaker the main body the less strong its outposts, and consequently the nearer to camp; otherwise the picquets might be destroyed before they could effect their retreat.

Armies or corps push forward their outposts, sometimes six or eight miles; and this distance diminishes in proportion with divisions, brigades, regiments, etc.

Where videttes, for instance, are four miles from the camp, their reserves should be half way between the two, that is, at two miles; the outlying picquets half way between them and the videttes (viz. one mile); and the non-commissioned officer's post half way between the picquet and the videttes. These distances diminish, as already stated, according to the strength of the detachments.

With small bodies of troops the number detached for outpost duty is generally in the proportion of from one-third to one-fourth of their strength; with large bodies of troops a fifth or sixth only.

The nature of the country has much to do with the number of men employed. In open ground few videttes can see a long way. In intersected country, where gardens, thickets, rows of trees, heights, and other obstacles intercept the view, both videttes and picquets must be placed closer together, and therefore require to be more numerous.

Cavalry forming a chain of outposts in an open country can place videttes by day at from six hundred to one thousand paces, for they can see each other at that distance, and also hear a shot fired.

A regiment of three hundred horses on outpost duty would keep one hundred in reserve; one hundred form the support, one hundred the line of outposts, of which fifty form the outlying picquet; the remaining fifty are detached as a post, from which sixteen videttes are supplied for the chain. These, at seven hundred paces from each other, enable you to cover eleven thousand two hundred paces of ground, or about six miles; equal to what it is calculated to require one thousand infantry soldiers to do.

But all these things depend so much on circumstances that no exact rules can be laid down, and much must always be left to the good sense of the officer in command.

Thus cavalry videttes may be placed sometimes at greater distances, and with good effect; whereas infantry videttes are generally at one hundred yards, and sometimes, in a close country, they are drawn together to within forty or fifty yards of each other. Both must, of course, lessen their circle and draw closer together at night.

When troops encamp for any length of time in one place, in addition to the common measures for security which are always adopted, they strengthen their outposts by abattis, redoubts, etc., and their chains of outposts by fortifying farm-houses, barricading villages, bridges, defiles, and in other ways render all possible approaches tn the camp both difficult and dangerous.

The outposts are under the command of some one specially appointed for this service, and to him all reports are made, and from him all orders received and carried out.*

Having thus endeavoured to convey an idea of the general use and employment of troops on outpost duty, I shall now take a squadron, and, considering myself detached with it on picquet, proceed accordingly.

The squadron is formed, and consists of ninety men and horses. I collect the nominal rolls of the men from the non-commissioned officers, then inspect the ammunition and fire-arms, and order the men to load. Finding they have neither forage nor provisions, I report the deficiency in writing; and having ascertained who commands the outposts, and where the reports are to be forwarded, I form an advanced guard with flanking parties and a rear guard and march off to the ground I am directed to take up.

Whilst on the march I ride forward and examine the country, taking notes of all places where I can make a stand in case of attack. Places where troops can act only on a narrow front, and where the flanks are inaccessible to the enemy, can be held by a small party (determined to do its duty) against almost any odds. For instance, a bridge, the road through a thick wood, a causeway running through marshy ground, etc.

---

*The troops employed on this duty form a curtain behind which the movements of the army are carried on, concealed from the view and knowledge of the enemy, and by vigilance and bravery they must endeavour to render that curtain quite impenetrable.

On arriving on the ground, I choose a spot for the picquet close to the main road and behind a bridge, ravine, or wood, so that I can make a stand at once when attacked.

I then divide the squadron into three equal parties of thirty men each. One of these, No. 1, moves off to the front, whilst the other two dismount; No. 2 remains in readiness; whilst No. 3 unbridles. A dismounted sentry is at once placed a little in advance of the picquet to give notice of anything that takes place in front.

No. 1, having advanced to where the non-commissioned officer's post is to be established, "halts." I take one-third of their number to the front to be posted as videttes.

To fix upon the requisite number, I ride up to the high ground and observe the number of roads, etc.; then give directions to the non-commissioned officers as to where to place them; and whilst this is being done, I make a little sketch, marking roads, rivers, bridges and fords, morasses, hollow roads, woods, towns, villages, and their distances. Having thus acquired a knowledge of the country, I fix upon the spot where the videttes, post, and picquet are to be placed at night, and communicate my orders to the non-commissioned officer in command of the post.

The non-commissioned officer of the post reports that on comparing the strength of his post with his videttes he has three men for each relief, and some men to spare. His videttes will therefore be relieved every two hours; the spare men are sent to patrol, or employed as extra sentries, during the night, and his post is relieved every six hours: that is, No. 1 is followed by No. 2, and No. 2 having been six hours on the outpost is relieved by No. 3 from the picquet, and so on till the picquet is relieved. The post now dismounts, and places a dismounted sentry in front to watch the videttes.

Patrols are sent from the picquet to communicate with the picquets on either flank, and at uncertain hours between reliefs to visit the videttes, and patrol along the front and flanks; before daylight a patrol goes to the front and on to the highest ground, then it awaits daylight and watches for signs of the enemy.

General Rules by Day

If it should be found necessary to have two or more posts instead of one in front of the picquet, in consequence of branch roads or heights interrupting a view of the videttes, the non-commissioned officer's party is divided into so many parts, each relieving their own videttes, but all other arrangements remain as before.

If the picquet is a small one, and the country to be watched is extensive, then the picquet must be divided into two parties, each relieving the other. In such cases the picquet should be relieved oftener from the main body. The feeding of horses on picquet takes place by divisions, one hour after the return of the morning patrol, at noon, and an hour before dark. When the men go to water their horses, they must bridle them up, and take everything with them.

The men must not be allowed to go into houses or villages in the neighbourhood.

The principal object of videttes is to notice the advance of an enemy; therefore they should be so posted as to give the widest scope for their observations which the nature of the ground will admit of. In order to spare men and horses, no more videttes than necessary are to be out.

In a fog, videttes are taken off the heights and drawn closer together, much in the same way as at night.

When a vidette discovers anything on the side of the enemy, he moves his horse in a circle at a walk. The officer, or non-commissioned officer, from the post, at once rides out to see what it is. Occurrences of all kinds, relating to movements of the enemy, are reported by the officer in command of the picquet.

If videttes see troops marching towards them, but yet far off, they ride the circle at a trot. The officer goes out as before.

If the enemy's troops are advancing rapidly, and at no great distance, for instance a mile, then the videttes circle at a gallop, and the officer mounts his picquet and at once advances.

If the enemy is so near as to drive in the videttes they fire off their carbines to give the alarm.

All strangers or deserters must be made to halt, lay down their arms, and advance singly to the videttes. When many come together, a party from the post or picquet goes out to meet them.

Picquets should not be pushed across bridges or causeways, unless their reserves or other troops are in position close to them.

A picquet should not be posted within musket-shot of any covered ground, as, for instance, near the edge of a wood, in dells or defiles where passes lead to the rear which are not occupied.

Picquets should be covered from the enemy's view.

In case of necessity, when you have not sufficient troops to organize a regular chain of outposts, the best plan is to leave out the supports and picquets, and merely establish small posts with videttes beyond. The main body, being then considered as the support, must be ready to turn out at a moment's notice.

General Rules by Night

If the enemy is near, no fires are to he lighted. As soon as the videttes can no longer see each other distinctly, it is time to take up a position for the night. Do not post videttes near rushing water, mills, or near anything where there is noise; for by night they must depend upon their sense of hearing more than their sight. When near the enemy part of the post and picquet must be mounted, the mounted men being pushed forward to give the dismounted ones time to get on their horses if attacked.

At night the videttes should be relieved every hour. They are close to the posts, and this can be easily done; the reliefs going round do the duty of patrols at the same time. When any part of a picquet or post is mounted, no dismounted sentry is required in their front.

Double videttes are always best, for one man can bring in the report of any occurrence to the post without loss of time. Videttes are taken off the hills or heights, and placed on the roads behind fords, bridges, or ravines. In clear moonshine they should be concealed in the shadow of a tree or bush. If they hear a suspicious noise, one vidette rides in and reports it; no one, whether deserter or otherwise, is allowed to approach too close at night, but is ordered to dismount and wait till the relief or patrol comes round.

If attacked and driven in, the videttes must not retire straight to their post, but some hundred yards to the right or left of it.

The posts and picquet have thus time to attack the enemy's flank or rear.

You do not skirmish by night, but hold every defensible position, and fire as much as possible to alarm the camp.

Relieving another Picquet.—Take over all written instructions, and write down the verbal orders received. Copy the sketch of the country made by your predecessor. Ask to whom reports are to be made, and what improvements are recommended for the security of the picquet.

Have all unknown roads, as well as the spot where the picquet and posts will stand at night, pointed out to you by men of the old picquet. Ride round with the officer of the old picquet to the relief of the videttes, and hear the instructions given.

All these things must be punctually attended to, or the original orders would vary greatly after a few reliefs had taken place.

Skirmishing.—The skirmishing of cavalry, as now carried on, is notoriously ineffective. "It is certain," says Warnery, "that hussars will sometimes keep up a scattered fire for a whole day with very little effect." I have heard it stated that cavalry was known to skirmish for whole days in the Peninsula without having a single man or horse killed, or even wounded.

A man in pursuit of another, or being himself pursued and riding at speed, may fire with some chance of hitting his mark, but he is not likely to hit anything while moving round in a circle, as our skirmishers do— trying to stop their horses for a moment, so that they may fire. This ought to be remedied.

In an intersected country let every second or third man dismount, and pick off the enemy with his rifle.

In an open country, if you are required to keep the enemy at a certain distance from your columns, ride out and occupy the line, but do not waste your ammunition. If the enemy keep at a respectful distance, well and good; if they close, let your men charge sword in hand, and decide the matter.

Some of the Sikh Horsemen often dismounted in battle, and if you charged they proved more formidable with these dismounted men amongst them than when they were all in the saddle. If you rode at the dismounted man, the mounted trooper would interpose, and, while you were engaged with him, the man on foot would quickly shoot you down, or knock over your horse.

Are there not easy means of securing and making something of the horse from which our dragoon or hussar dismounts? If the collar-chain had a hook, the right file would fasten the horse of the left file to his own; the right file would thus have the full use of both his arms, and, if attacked, the led horse, instead of being in his way, would effectually protect his weak side (the near one) from an assailant.

There can be, therefore, no great danger in pursuing this plan; when the enemy is at a distance it is the only way to do execution with the rifle.

In intersected ground it would give a decided advantage, and a few active fellows, thus trained, and good shots, would at any time inflict heavy loss on columns of heavy cavalry, whose pistols would be but a slight protection against the rifle.

It would at least make the men active, to practise skirmishing in the following manner at squad drill:— On the signal to skirmish, the right files close to the left, to hold the horses; the left files spring to the ground, unsling their rifles, move a few yards to the front, or further according to the ground, and open fire till the advance or retreat, or right or left, is sounded; they then mount and take the place of the right files, who dismount in their turn on the next signal to skirmish.

When retiring, take advantage of all natural obstacles—get quickly behind them, and occupy them with dismounted men.

It is no easy thing to clear even a very small ditch in the face of a fire of small arms; the enemy must try to dislodge you on foot (unless he can turn the obstacle), which will give your men time to inflict loss upon him, and then continue the retreat.

In an open country always retire by alternate ranks with drawn swords, and when the enemy presses closely turn upon him and charge.

Squads should practise against each other—advancing, retiring, etc.; practising the attack and defence of bridges and defiles, drawing together in the advance, to be ready to charge, and scattering at the signal to skirmish.

They should charge one another, each man circling round his opponent to the right at a gallop, crossing swords, then returning again to their places.

The reserves always advance to the support.

"The Cossack skirmishers often close together for a charge, but disperse in retreating.

"Our skirmishers almost invariably do exactly the contrary, whatever may be the facilities offered. The Cossacks are right, and we are wrong. In fact, how often have we not seen our troopers, when retiring, rush one upon another in such a manner as to deprive them individually of the use of their arms, hampering and arresting their progress! and this, by still farther retarding that of men not so well mounted as themselves and keeping them to the rear, has been the means of causing these poor wretches to be cut down or taken prisoners, their backs answering as a shield to their guilty comrades.

"A too crowded retreat is always put to the sword, for two very simple reasons; the first is, that the troopers who are retreating, by crowding excessively together, paralyze their individual means of defence, and render both the halt and the wheel about impossible; and the second is, that the attacking trooper, who has only one object in sight, and who is not at all harassed upon his flanks, hurls himself forward with all his impulsive power and all his audacity upon this unresisting mass, which he hacks and hews in perfect security, and can drive before him as far as he chooses.

"This is not the case in a retreat in which the men scatter as they go; a man who retreats in this fashion preserves all his defensive power; he is equal in all points to the attacking party, who takes care not to rush recklessly upon him, because his flanks are threatened, his attention diverted, and the danger equal on both sides. A retreat conducted in this manner is never so vigorously pressed, nor so far pursued. The slowest horses will perform it as well as the swiftest; it disturbs the attacking party, stops him short just when he might have obtained the advantage, and it can wheel round and resume the offensive. Polybius tells us, in his description of the passage of the Trebia, 'Nevertheless Sempronius caused the retreat to be sounded, in order to recall his cavalry, which did not know how to manoeuvre against the enemy in its front. In fact, it had to deal with the Numidians, whose custom was to retreat scattered in different directions, and to return vigorously to the charge when the enemy least expected it.'"*

At Stockholm, on the 29th June, 1852, I saw the regiment of Swedish Horse Guards, under Count Stedingk—a fine body of men, but rather under mounted.

They had four squadrons, each of about seventy-five horses, divided into three divisions, which they call troops. The squadron-leader alone was in front, an officer on each flank, and one in the rear: they had no markers, and no one out of the ranks.

They skirmished by squadrons against each other, frequently sounding a charge, when the men, seizing their swords, galloped towards each other, each man riding round his opponent at a gallop and crossing swords; the reserves advancing to the support.

On a signal to advance at a gallop, the skirmishers drew together towards the centre of their divisions. and, after the charge, resumed their line.

The regiment went through a field-day.

They change front on the move in open columns of squadrons. They always form a close column in rear of the squadron named. Their open column is one of squadrons generally. They advance in open column from any named squadron, and, when forming line, the squadron which has lost its place in line resumes it on re-forming, and takes the shortest road to get there.

I was much struck with the deep silence that reigned throughout. The colonel's word of command was followed by a caution from the squadron-leader only; the trumpet sounded and the movement was executed. There was no noise, no shouting, no hurrying. If a halt was sounded, they stood motionless; all hung on the word of command; not a move was perceptible; no dressing up. no reining back, no closing in; not a word was spoken!

*De Brack.

Their working in the field was as fast as the power of their horses would admit of. In coming into line they halt the squadrons in rear of the line of dressing. The squadron-leader and the officers on the flanks of the squadron move up to the alignement, and then the squadron dresses up between the officers.

Their charges were excellent and made at their utmost speed (I rode with them to try it): the flanks were always well closed up; no flying out was perceptible; and if they opened in the least during the advance, they closed up gradually again. There were no breaks in the line at any time, and they halted at once and all together at the signal. They have, instead of a cloak, a good thick great-coat with sleeves, in which they can ride, fight, or do anything.

For parade purposes they have a slow canter: when marching past by squadrons, and the reviewing officer is on the right, they lead with the right leg; when on the left, then the horses lead off with the left leg.

This is one of the best regiments of foreign cavalry I have ever seen.

I conclude these remarks on skirmishing with extracts from the letter of a distinguished foreign cavalry officer:—

"During great part of the last war against the French I was attached to the Cossacks of the Don. These men were at that time but little accustomed to the use of fire-arms. Whilst advancing into Western Europe the advantages of fire-arms became apparent; more particularly when acting in intersected and difficult ground: and the Cossacks managed to arm themselves with French infantry muskets which they picked up on the field. Then originated amongst them the practice of dismounting by turns where the ground was favourable, and thus engaging the enemy in skirmishing order. I have myself seen them in this way beat cavalry very superior to them in numbers, and infantry also, when either the cavalry or the infantry attempted to attack them singly. In such cases the infantry soldiers opposed to them were afraid of the mounted men, who stuck close to their dismounted comrades with the led horses; and these dismounted men were ready to jump into the saddle at any moment, and rush upon the enemy, if they gave way or were driven from their cover.

"*To this manner of skirmishing I attribute entirely the success of these Cossacks during the campaigns on the Elbe and the Rhine, and the decided superiority they acquired over the enemy's cavalry in all outpost work and detached warfare.*

(Signed) "H. v. Ganzauge

"Captain in the 2nd Regiment of Prussian Lancers of the Guard."

12

THE CHARGE, THE PURSUIT, AND THE RALLY

"Une charge en ligne n'est réellement qu'une suite rapide de charges successives, dont les braves forment les point saillants; ce sont ces causes qui rendent le succès des charges si incertain, et qui doivent faire éviter d'en entreprendre sur de grands fronts."—C. Jacquinot de Presle

T HE CHARGE MUST BE DECIDED promptly, and executed vigorously; always made and carried out at speed.

The first object is to break through and disorder the enemy's array, then make use of the sword to complete his discomfiture.

Powerful horses urged to their utmost speed, their heads kept straight and well together, will seldom fail to attain the first object in view: sharp swords, individual prowess and skill do the rest.

Officers must bear in mind that, however successful a brave and determined body of horsemen may be, there is a limit to everything. The horses must in time get blown, the men tired out, the squadrons scattered; they are then at the mercy of any body of fresh horsemen.

Reserves must always be at hand to follow up steadily any success achieved, or, in case the first line is brought back, which is sure to happen sooner or later, to fall upon the pursuing enemy, and give the fugitives time to re-form.

The first line will never think of turning so long as every man in it feels that the reserves are close at hand to back him up. Thus, in every case, the reserves should follow closely and be ready to act whenever and wherever their action is required. When a charge has once begun, carry it out whatever may be the odds that suddenly present themselves against you. As your first line moves forward, your reserves, distributed in different échellons, take up in succession the positions which have been carried.

Innumerable reverses are attributable to the neglect of these rules about reserves.

In an attack upon cavalry formed and advancing to meet you, spare your horses and husband their resources for the hour of need. Bring them up at a trot to within one hundred and fifty yards, then sound the gallop, and immediately afterwards the charge: thus brought up close to the enemy fresh and well in hand, you hurl them upon him in close array with irresistible speed.

Should the enemy begin galloping earlier, you must do the same, for lines approaching each other close very quickly; and therefore be careful not to be caught by the enemy's cavalry before you have put on your speed to meet him.

Artillery in an emergency may urge their horse to the utmost, because when they come into action their horses are at rest; with cavalry it is the reverse, for it is *then* that the horses are required to exert themselves most.

If you meet the enemy's cavalry with blown horses, you are pretty sure of being thrown; but even should success attend the first rush, that success would be useless, for it could not be followed up.

In heavy ground bring your men slowly as near as possible to the point of attack.

If the enemy's cavalry is in column, or in the act of forming, sound at once a gallop, and try to overthrow him before he has completed his array.

Against infantry squares formed to receive you, begin to gallop as soon as the first fire has been delivered, but do not let the men rush forward at speed until they are within fifty yards of the square.

Charges of cavalry on a large scale, against masses of troops of all arms, should be carried out with the greatest impetuosity and speed; no time must be allowed to the enemy to prepare.

No distance can be laid down at which to charge, it depends on so many different circumstances. When the ground is favourable and your horses are in good condition, you can strike into a gallop sooner; but the burst, the charge itself, must always be reserved till within 50 yards, for in that distance no horse, however bad, can be left behind, nor is there time to scatter, and they fall upon the enemy with the greatest effect.

The Pursuit and Rally.—If you have succeeded in overthrowing the enemy's line, your own will be in disorder. The mêlée which ensues, soon, however, turns into a pursuit, and this affords the opportunity of destroying those who have turned; for the charge and mêlée do not last long enough to inflict or sustain a heavy loss in men or horses.

If a defile is in the rear of the enemy, the first duty of the officers pursuing is to dash on with some well-mounted men to intercept and cut off the line of retreat. When this has succeeded, then make prisoners.

The *pursuit* must be kept up with vigour. Each man singles out his foe, and woe to those who are pursued by men with better horses, whose superior speed brings the avenging sword upon them; whose better blood and condition give them endurance to continue the pursuit!

In such circumstances it is difficult to recall the men: they are maddened with the excitement of the fight, and intent only on cutting down the unfortunate fugitives in front of them. This is not the time to stay the slaughter, but watch over the safety of the pursuers with your cavalry reserves till the flying enemy is entirely dispersed.

Then rally. If you try to re-form line and pursue with flank troops or reserves, you only give your enemy the opportunity of riding through the intervals of his second line and there re-forming.

The line rallies in the order the attack began.

Each troop rallies for itself, and, when formed, is led into line.

Second Line.—If the enemy have a second line behind the first one, drive the first one upon it, giving them no time to choose the road. Dash into the second line pell-mell with the fugitives, which is sure to disorder it. It will probably be carried off with the stream; if not, detachments from your reserves must do the needful.

Should the enemy's second line be formed on the flanks, then you must rally from the pursuit as quick as possible, and, together with your second line or reserves, charge them instantly.

Time is here of great moment, and each troop, when rallied, should at once be led against the foe. Attack in succession, in échellon, or anyhow, the great object being to throw the second line into disorder before the first one can rally again.

Once a line of cavalry hurled against the enemy, all orders from the commanding-officer must necessarily cease for a time. The men of each troop look to their leader; the officers on the flanks keep them as much as possible together in order to concentrate their efforts round their captain.

The captain must act with his troop, without hesitation, wherever opportunity offers: he must never remain idle nor wait for orders whilst an enemy is near.

13

Detached Service

E VENTS IN WAR ARE SO VARIED, that to lay down rules for all cases would be quite impossible; still it may be useful to point out the application of a few general rules meant to assist officers in command of detachments of cavalry in the field.

1st. Unless taken by surprise never engage the enemy without having reconnoitred his strength and dispositions.

2nd. Never retire before an enemy without taking the same precaution. To do the first proves temerity and want of head; to do the second shows timidity and want of heart.

Reconnoitring a superior force of cavalry requires judgment, for it is often attended with danger; but infantry may be reconnoitred in perfect security, for they cannot pursue.

3rd. Hesitation is always dangerous in war; decide promptly one way or other. Better a wrong decision than a delay which will not allow you to decide at all, and which must expose you to the will or decision of your foe.

4th. When you can take an enemy by surprise, always do so. By stratagem you attain your object more easily and with less loss of life than by open force.

5th. A reserve, however small, appearing at a critical moment in a charge, will generally decide it; therefore, however small your detachment, always keep a few men in reserve, and, if possible, out of sight of the enemy: their sudden and unexpected appearance will then have more effect. Even at the risk of repetition and a little tedium, for the benefit of young cavalry officers I dwell upon these vital points.

Even when supported by infantry, behind whom you can retire in case of defeat, always keep a reserve (unless you have not sufficient men to charge with): the probabilities are that the enemy has a reserve, which, when you are engaged with him, will attack you in rear or flank and cause your defeat, unless you have some men ready to attack him in the same way.

6th. Always pass a defile quickly. When through, attack at once with the leading detachment, and keep the outlet clear for those following.

Form the reserve at the entrance of the defile; for if the troops are driven back, it stops the pursuers and covers the retreat.

Occupy half the road only, and make the men keep the near side of it to prevent the troops getting jammed up when attacked, and to keep that side clear on which the cavalry soldier makes use of his sword (the right).

Before entering a defile reconnoitre it. Wooded places and defiles of no great length are reconnoitred by making a man of the advanced guard gallop through, followed at some distance by another single horseman. If there is no direct impediment, the advanced guard then gallop through and keep the outlet till the main body has passed out.

When a cavalry detachment without infantry has no choice, but is obliged to pass a defile known to be in possession of the enemy, their best chance Is to attempt it at night and at a gallop. If a barricade has been raised, the advanced party give notice, and a few dismounted men endeavour to remove it. If the obstruction is of such a nature as to be impassable on horseback and impossible to remove, the party will have to turn back.

When acting with infantry, these reconnoitre the defile, for it would be dangerous for the cavalry to find an enemy's infantry posted there on their flanks, and out of reach of their swords.

When a defile is occupied the infantry try to force it, and the cavalry endeavour to turn it.

When, in passing through a defile, you are suddenly attacked by cavalry, charge at once; if you turn you are lost! You cannot get away, and are sure to be cut down; whereas by charging you place yourself on equal terms with the foe, who, whatever his numbers, cannot bring more men into line than the breadth of the road will admit of, and you can do the same.

If you have infantry with you in a defile, and you meet the enemy, let the infantry try to gain his flanks while you charge him in front.

7th. When in pursuit you come up to the enemy at the entrance of a defile or village where he has made dispositions to receive you, then leave a few men to watch him, and ride round the village, or turn the defile and cut in on his line of retreat.

8th. Always attack at once any working parties left by an enemy to destroy bridges or roads, etc.

9th. In covering a retreat through a defile endeavour to stop the enemy's cavalry by upsetting waggons or by otherwise incumbering the road; and defend these barricades with a few dismounted men.

If you have no means of barricading, form at the outlet of the defile with a party on the flank, and charge the enemy in front and flank simultaneously the moment he attempts to force his way out.

10th. Rivers should not stop cavalry, unless the banks be so steep as to afford no footing for the horses to land. There are plenty of examples of cavalry swimming rivers without loss. For instance—the night before the battle of Hastembech 300 horse, with as many foot soldiers, were detached from the camp of the Duke of Broglie and swam the Weser, the foot-soldiers holding on to the horses' manes. Again, 24 squadrons of Austrian cuirassiers swam the river Main on the 3rd of September, 1796, leaving the bridge for the use of the infantry.

Cavalry should swim by ranks in line, going in abreast to break the current. The men ought to look to the opposite bank, not at the water; for if they do so they are very apt, insensibly, to follow the current. They hold by the mane, and only use the snaffle rein to direct the horse: both reins should be tied up short to prevent the horses putting their feet into them when swimming.

When boats are at hand and no enemy in the near neighbourhood, the men's kits and arms should be taken across in the boats, to keep them dry and make it easier for the horses to swim.

11th. Cavalry never surrenders, under any circumstances, in the open field, but must always attempt to cut its way through, or, by scattering, elude pursuit.

If an enemy is expected to cross a river in your front, do not scatter the cavalry along its banks, but concentrate it at some distance from the river in different points. It can then move up in force quickly with its horse artillery to any place at which the enemy attempts the passage.

In the attack and defence of entrenchments, cavalry in the defences is kept out of reach of shot, and, when the enemy is in the act of storming, it gallops out of the entrenchments and charges them in flank,—going out at the right and re-entering the entrenchments by the left—or *vice versa*. When this is not feasible it waits inside till the enemy have effected an entrance, and then charges to drive them out. When the outlets are narrow and only admit of files, it does the same, because in filing out it might be taken in flank, and if defeated it would not escape through so narrow an entrance without sustaining heavy loss.

In the attack, the cavalry is masked, when possible, behind the infantry columns, ready to meet a sortie on the part of the garrison, or to profit by any means of entering the entrenchments which may be secured for it by the infantry. It may be sent to turn the entrenchments, or otherwise assist in their capture; but it is out of place for cavalry to attempt to take a leading part. A division of French cuirassiers that tried to ride into the entrenchments at Wagram were nearly destroyed without doing any good.

In sieges, when a place is to be invested by a rapid march, the cavalry occupy all the surrounding passes and villages in the neighbourhood, and hold them till the infantry come up. They then escort convoys, bring in provisions to camp, etc.

When with the besieged, they are of use in making sorties, forays, etc.; and when the outworks at last fall into the hands of the enemy, the dragoons either serve the guns or retire from the fortress, and throw themselves into the enemy's rear.

To effect their retreat the garrison make a sally at night, and the cavalry escape in the dark.

14

Cavalry Charges Against Infantry

"Zu allen Zeiten, wo die Kunst verfiel,
Verfiel Sie durch die Künstler."—Schiller

I HAVE ALREADY SPOKEN BRIEFLY, of cavalry versus cavalry, and of cavalry against artillery. I have purposely reserved the matter contained in this chapter for the conclusion of my little work. It will show what cavalry has done, and what it may do again. I will not affirm, with Colonel Mitchell and some other military writers, that cavalry, if properly armed, mounted, and led, will, under every circumstance or combination of circumstances, break an infantry square —I will only say that it *may* frequently do so. Let no infantry officer take offence at my opinion. A greater respect for our own matchless foot-soldiers than that which I entertain cannot be felt by any man either in the service or out of it. But if I carry any of my propositions too far, let it be remembered that I am a *cavalry* officer.

Cavalry is now often reduced to play a secondary part in war, in consequence of the greater ease with which armies in the field change their positions. The infantry, no longer tied down to a battle-field chosen for its level ground, or a position covering their encampment, more lightly armed and equipped, move off during an engagement, if threatened by a powerful cavalry, and neutralize its effort by occupying enclosures, or getting into a hilly or otherwise difficult country.

Instead of forming long lines and open columns, they move in compact masses, or throw themselves into squares, which, supporting each other by a cross fire, and protected by artillery, often defy the efforts of cavalry. But can they always move in an open country, and laugh at the efforts of that arm, the thundering of whose approach was sufficient, formerly, to inspire them with awe?

If infantry squares are impregnable and artillery safe within the swoop of bold horsemen, then our cavalry is no longer indispensable to the efficiency and *safety* of an army.

Why then are such large sums expended by every nation in Europe to organise an efficient cavalry, and raise their numbers to nearly one-fifth of an army in the field? why this expense if cavalry is henceforward to be used only for outpost duty, for skirmishing, for carrying despatches and orders in the field?

Unfortunately, in our cavalry the general opinion is against the possibility of infantry squares being broken.

If troops believe a thing impossible, success is not to be expected. If cavalry are led to suppose that they cannot break an infantry square, they will add one more example to the frequency of such failures whenever they are next ordered to charge.

If to infantry soldiers you admit that they cannot stand exposed to the fire of artillery, such men will be always ripe for defeat.

Those whose duty it is to instruct the men under their command should instil confidence in the power of that arm to which they belong, and never allow soldiers to doubt of success when they do their duty manfully.

In the Seven Years' War, as much, if not a greater proportion of artillery was used than in more modern times; for instance, at Leuthen, Frederick the Great had between 33,000 and 34,000 men in the field, and 167 guns, a proportion of five guns to every thousand men; in the more recent campaigns the average has seldom exceeded three guns to 1000 men; but neither the numerous artillery nor well-drilled infantry stopped the cavalry in those days.

The numerous improvements effected in the artillery are in favour of the cavalry, not against their chances of success.

No infantry can now escape from its fire, and when subjected to it the better practice will throw them sooner into disorder, and make them ripe for the harvest of the sword.

Horse-artillery can move with almost equal speed and act in concert with cavalry, where formerly the cavalry must have acted alone. With such powerful assistance (under almost all circumstances), cavalry are surely more formidable than before, and with horse-artillery they must always destroy infantry, however good and tried it may be; for, supposing the cavalry alone can do nothing, the artillery can destroy them with its fire if they keep together, and if they attempt to deploy they must fall a prey to the horsemen. Both artillery and cavalry can keep out of range of the infantry's fire all the time.

Thus, if the improved tactics of infantry have given them an advantage over the cavalry, the latter have more than made up for it by the improvements of the artillery, which, by keeping company with them, can, by their fire, afford the cavalry those favourable moments at which to charge is to conquer.

An infantry square ought to be attacked with a front smaller than the side of the square charged.

If you outflank it, the men on the flanks ride round the sides, and those in the centre open out and follow suit.

The officers on the flanks must look to this, the cavalry put their horses to their utmost speed, ride home, and the square must go down.

Saddles will be emptied, horses killed and wounded, but no horse, unless he is shot through the brain, or has his legs broken, will fall; though stricken to the death, he will struggle through the charge.*

When the ground and other circumstances admit of it, a good plan to attack a square is, for two troops to charge, one the front side, the other the adjoining side, of the square; a third troop meanwhile forms opposite the angle of the square, the two sides of which are charged at

---

*For a horse must be very much wounded to make him fall upon the spot. "One without his rider, at Strigau, which had one of his hind feet carried away by a cannon-ball, joined the left of the squadron, where he remained with the others during all the battle, although we were several times dispersed; at the sound of the call he always fell into the same place, which was, without doubt, the same that he had before belonged to in the squadron.

"Another time, a cuirassier's horse fell, in the grand attack at the exercise at Breslau; the cuirassier got him up again and mounted him; at 300 paces he fell down dead. The late General Krokow, colonel of the regiment, had him opened, and it was found that the sword of the cuirassier had penetrated his heart a tenth of an inch. These facts prove that a horse is not easily to be brought down, unless a ball should break his skull."—General Warnery

the same moment by the first two, who having drawn the fire, the third troop rushes down, and is upon the square before they are aware of its approach.

A favourite manoeuvre of the great Frederick, and one executed with great success by his cavalry under Seidlitz at Zorndorf, under Marshal Gesler at Strigau, and under General Lüderitz at Kesseldorf, was to form a close column at some distance in front of the centre of a line of cavalry, and thus bear down on the enemy's infantry,

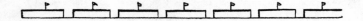

and attack them at speed. When they broke through the line, the two rear squadrons of the column wheeled outwards and rolled up the broken infantry, whilst the cavalry in line rode over them and followed the leading column to be ready to fall on the enemy's cavalry should it press forward to the rescue. The Prussian cavalry were ordered to shout and make as much noise as possible when attacking infantry to prevent their hearing the whistling of the bullets; but when charging cavalry the deepest silence was preserved, and all hung in breathless suspense on the word of command which was to hurl them simultaneously on the foe.

I will now adduce some instances of successful charges of cavalry against infantry squares, namely,—

At the BATTLE OF FRAUENSTADT the Swedish dragoons cut down the whole of the Saxon infantry formed in squares.

At HOHENFRIEDBERG the dragoon regiment of Baireuth rode over 21 battalions of infantry, took 4000 prisoners, 66 stands of colours, and five pieces of artillery.

ZORNDORF.—[See ante, page 19.]

Combat of AVESNE-LE-SEC.—3000 French infantry, with 20 guns, formed in squares to receive the charge of four Austrian cavalry regiments, under Prince Lichtenstein and Count Belgarde, and were overthrown at the first onset: 2000 men, five stands of colours, and 20 guns fell into the hands of the victors; the remainder of the French were cut down, with the exception of a few hundred stragglers, who reached Cambray and Bouchain.

VILLIERS-EN-COUCHE.—The French, 15,000 strong, were defeated by ten British and four Austrian squadrons: part of this force dispersed the French cavalry, whilst two British and two Austrian squadrons broke through the French square, killed 900, and took 400 men prisoners, together with five pieces of cannon.

CÂTEAU CAMBRESIS.—One regiment of Austrian cuirassiers and nine British squadrons defeated General Chappuis' army, 27,000 strong, inflicting a loss of 3000 men, and capturing 22 guns and 29 ammunition-waggons.

AFFAIRS OF EDESHEIM AND KAISERLAUTERN.—Marshal Blücher defeated the French at Edesheim with two cavalry regiments; and at Kaiserlautern charged 600 French infantry soldiers with 80 hussars: though they were prepared to receive him, he broke in, killed, wounded, or captured the whole party.

NORDLINGEN.—On the retreat of the Austrian cavalry from Ulm, in the year 1800, they broke and defeated three regiments of French infantry, belonging to the division of General Montrichard.

AUSTERLITZ.—The Russian cavalry broke the squares of French infantry formed by the brigade of General Schinner, division Vandamme of Soult's corps; and a regiment of their lancers broke the square formed by the fourth regiment of the line, and captured their eagle.

AUERSTADT.—The Prussian dragoon regiment of Irving destroyed a square of French infantry, which stood firmly to the last, and gave them a volley at fifteen paces which brought down nine officers and many men; but the dragoons were not to be stopped; they rushed in and cut them to pieces.

WALTERSDORF, the 5th of February, 1807.——The French cavalry overtook the Prussian rearguard, consisting of five battalions, ten squadrons, and a battery of horse-artillery. They, charged and overthrew the cavalry, and, after a gallant resistance, nearly destroyed the whole of the infantry. The remains of some battalions alone were saved by the charge of the Prussian hussars of Prittwitz.

At the BATTLE OF PULTUSK the Russian General Koschin charged and overthrew the right wing of the French.

On the 5th of May, 1809, the Prussian Major Schill, with 600 hussars, attacked 1800 French infantry, who were posted with two guns in broken ground. Their commander, General Michaud, wrote as follows to the minister at war, Eblé:——"Ces hussards ne se battent pas comme des soldats ordinaires, mais comme des enragés: ayant rompu et sabré mes quarrés, ils firent les restes prisonniers. Venez à mon secour," etc.

ACTION OF GARCI HERNANDEZ, 23rd July, 1812.——Three French squares were broken by the King's German Legion.

COMBAT OF THE GOERDE, 16fh September, 1813.——Here a corps of 9000 French with fifteen pieces of artillery, partly posted in broken ground and partly in the open, after defying the attacks of the Cossacks and the cavalry of the Russian-German Legion, were charged by the 3rd Hanoverian Hussars, and those squares which were in the open were completely defeated with the loss of their guns.

AMBUSCADE OF HAYNAU, 26th of April, 1813.——Here General Maison's division of the victorious French army, eight battalions, with eighteen pieces of artillery, were ridden over and captured by Colonel Dolfs at the head of twenty Prussian squadrons. [See ante, page 33.]

It is impossible for the infantry soldier physically to resist the power of a horse when at speed; that the horse will face both fire and bayonet is proved by the many examples given.

The horse often feels the man's unwillingness to go on, and turns, but then it is in obedience to the bridle-hand.

The following is a translation from Berenhorst's *Betrachtungen über Kriegskunst*. His reputation as a writer on military matters, and his personal experience in war, must entitle his views to some consideration:—

"Against cavalry, it is the rule for infantry to fire steadily by word of command, and not to begin too soon. Here again it is assumed that the cavalry will turn. All regulations are silent as to what is to be done when they do not turn after the last shot has been fired and the horses are on the bayonets.

"To demonstrate the matter better, we will examine, measure out, and calculate the chances of a charge of cavalry against infantry, conducted according to rule.

"Let one-sixth of the horses be shot down (the riders are not taken into consideration), this does not stop the advance of the remainder. Suppose the infantry in the situation as above, for which no instructions are given, namely, the volley has been delivered and the muskets are brought to the charge.

"The second and third rank may have their muskets at the charge, or be busy loading; but the front rank have their muskets thrown forward; the right hand grasps the small of the stock; in this position the musket and bayonet reach only three feet beyond the man's elbow. Is the infantry soldier now to aim at the dragoon, or his horse? He cannot reach the man: it is four feet from the horse's nose to the man's belly, and three and a half from the horse's forehead to the man's breast.*

"The man is further protected by the head and neck of his horse,† and, if the infantry soldier tries to thrust at him, he comes in contact with the horse and is thrown down.

"But let us admit an impossibility. Every bayonet has been buried in the stomach or breast of the horsemen; still, the horses alone will break the ranks of the infantry.

"The infantry soldier can only try and aim his bayonet at the horse's breast, and let him spit himself like a wild boar. In this case he must hit the heart to kill him, for any other wound would he quite useless at the moment; and even reaching the horse's heart cannot save him, for the

*Berenhorst.—This measurement is taken on horses of a small stamp.

†The horse's neck and the soldier's musket act like two swords crossed: if the one turns the point of the other from the straight line the thrust goes off sideways.—Berenhorst.

horse, with his great weight and the impulse of his speed, will dash the whole rank to the ground in his fall.

"Infantry can therefore depend only on its fire: it has time to deliver two volleys, no more, and therewith barely the power to knock over every sixth horse.

"Experience shows that the effect of musketry is very trifling at more than 300 yards; within that distance it would not be advisable to try more than two discharges.

"The calculation is that the fusileer can discharge his piece five times in the minute, thus for each shot 12 seconds of time are required.*

"The cavalry soldier will pass over 600 paces in 30 seconds, to each 100 paces 5 seconds.

"If the fusileer delivers his first fire at 300 paces, and 12 seconds later the second, he has only 3 seconds left, cannot load again, and will be ridden over in the interim.

"A battalion which delivers its fire at 60 paces, and wants to reload, is ripe for mowing down. As 60 yards is quite far enough to miss at, particularly when the rushing in of the cavalry shakes the earth and men's nerves at the same time, it would be far better to order one discharge only at 30 paces, and bring the bayonets to the charge without attempting to reload.

"Then the infantry are still in the position already alluded to, and in which the superiority of the cavalry has been demonstrated.†

"In speculating, however, on this subject, the advantage is entirely on the side of the infantry. Assume that each file fire true, and thus send their bullets into the breasts of the front-rank horses; those of the rear rank fall over them, and it is all up with the cavalry; the men are

*Note by Berenhorst.—"We wish to he generous, for with ball cartridge the soldier could not reload under 15 seconds."

†By the author.—It must be a very fast horse to go over 600 paces in 30 seconds. Cavalry could not well do it under 40 seconds. With the present musket the soldier could not deliver more than two volleys against an advancing line, for it would be useless to fire at more than 160 yards. The long-range gun is therefore not to be despised: and the needle gun of the Prussians, which can be loaded quicker and fired with greater accuracy, and with these advantages combines the long range, is a formidable projectile in the hands of a steady soldier!

stretched on the ground amongst a whole line of dead and dying horses, and all they can do is to pick themselves up and surrender.

"In the battles of civilized nations amongst the too sober warriors of the west, the probabilities are that the horsemen will gallop slowly; instead of plying their spurs within the last 300 or 400 paces, they will pull at the reins, and when the bullets begin to whistle they will turn and gallop to the rear by whole squadrons.*

"The infantry will fire too soon, fire badly, lose its presence of mind, and sometimes its order; both sides much alike.

"The success or failure will depend on those unlooked-for circumstances which in war generally decide between victory and defeat.

"Still, on carefully comparing the chances, the balance is in favour of the horsemen: the misfortune is, that they are seldom anxious to avail themselves of it. The cavalry of Charles XII and Frederick the Great alone used their advantage to some purpose."

At the battle of the Pyramids many single Mameluke horsemen dashed right through the French squares, and out at the other side. If only twenty of them, with the same determination evinced by these individual horsemen, had charged the French squares together, they must have succeeded; but they exhausted their horses in irregular charges, each man coming on in his turn to be shot down, till at last their courage was tamed by repeated failure, and they fled the field.†

In the pursuit, after the battle of Salamanca, at Garci Hernandez, where the English cavalry (German legion) rode over three squares of French infantry, five dragoons actually charged one square by themselves, broke in, and two of them cut their way through and got out at the other side.

At the battle of Aliwal a squadron of the 16th Lancers, under Captain Pearson, rode through the Sikh infantry: the gallant leader, dashing in alone, went through them first.

Captain Bere, of the same regiment, with a squadron, went right through a Sikh square, wheeling about and re-entering it again.

---

*Note by Berenhorst—"The author saw this happen himself."
†*Geschichte der Spanischen Monarchie, v. 1810 bis 1823*, v. Obersten v. Schesseler.

But our infantry have stood unconquered against the best cavalry of France in many a bloody field; they marched against them, and into the midst of them, at Quatre Bras, under General Picton.

The Sikh infantry were not wanting either in discipline or valour; they also at Aliwal advanced against the British cavalry.*

But infantry squares that stood firm and undaunted, that delivered their fire with good effect, have nevertheless been overthrown and totally destroyed by the charge of a few gallant horsemen.

Contrast the two following instances, which show that the safety of the infantry does not depend upon the courage, upon the steady discipline and firm behaviour of the square, but rather, solely, on the forbearance of the cavalry.

1. In the expedition to Russia in 1812, whilst endeavouring to save the wounded at Mojaick, fifty of the light company of the 33rd (French) clambered up a height, the summit of which was occupied by the enemy's cavalry and artillery. The French army, halted under the walls of Mojaick, looked on with astonishment at this handful of men, who, dispersed over the unprotected declivity, annoyed thousands of the Russian cavalry. The consequence, which might have been anticipated, soon appeared; several Russian squadrons were seen in motion, who, the next moment, surrounded these brave men. They instantly formed a square; but they were too few, among so many horsemen, and in so vast a plain, and were soon lost to the view of the French army. Some smoke, which arose from the centre of the mass, prolonged the uncertainty; the anxiety lasted for some moments, when all at once the army gave a shout of admiration on seeing the Russian cavalry disperse, in order to escape the well-directed fire of that handful of heroes, who were thus left masters of a large field of battle, of which they barely occupied a few feet.†

2. At the siege of Trichinopoli, in the month of February, 1753, a company of British infantry held a fortified post at some distance from the town: they were attacked by a party of Mahratta horsemen, and beat

*Sir H. Smith's despatches.
†Segur, *Expéd. de Russie*.

them off with great loss. When relieved at their post, they had to march across the plain to return to cantonments. The Mahrattas were waiting for them. The English soldiers marched out full of confidence; they were repeatedly attacked, but formed in square, and, reserving their fire each time till they could deliver it with deadly effect, they strewed the plain with men and horses, and continued their march, taunting the cavalry, and daring them to come on.

The Mahrattas, much reduced in numbers, but still determined to have their revenge, formed in two ranks, the second at some distance behind the first; they then advanced steadily up to the bayonets; the English took a deliberate aim, poured upon them a most deadly fire, and down went the leading ranks, men and horses: the Mahrattas had drawn the fire, as had been agreed amongst them; but over their prostrate bodies rushed the rear rank like infuriated fiends, to avenge their fall, and, dashing in through the bayonets, these horsemen killed every man in the detachment.

Near Augsburg, and between that place and Sulzbach, the French 20th regiment of light infantry, formed in square, was attacked by some Austrian cavalry, and beat them off twice; but the third charge was successful; the Austrians rode over them, and killed or captured the whole regiment.

At Rivoli the French cavalry did great execution on the Austrian infantry. Near San Giovanni the Cossacks attacked the French infantry under General Dombrowski, with complete success, and destroyed several battalions.

At Wertingen, in 1805, Murat, at the head of three divisions of cavalry, surprised and defeated General Auffenberg's Austrian corps of nine battalions, four squadrons: the infantry was formed in square and made a gallant resistance. Many fell under the sword; 2000 prisoners, of whom fifty-two were officers, eight guns, and three standards, fell into the hands of the French.

Many more instances can be found by referring to the battles of Medellin, Ciudad-Real, Margaleff; Todendorf. The French say that their 7th and 9th regiments of chasseurs broke *three* squares of English infantry at the battle of Fuentes de Oñoro.

I now proceed to my last examples, taken from the Hungarian war, to show that neither the improved fire of artillery nor the long-range muskets could save the squares when the cavalry did their duty.

"On the 28th of December, 1848, at eight o'clock A.M., General Görgey's rear guard, consisting of two battalions of infantry and a small detachment of hussars, were attacked by the cavalry brigade (of the Austrians) under Colonel Ottinger, who, in spite of several discharges of musketry from the squares, continued their steady advance, charged, broke the first square completely (formed by the regiment of the line Preussen), and partly destroyed the second square, inflicting on us a loss of 300 to 400 killed and wounded, and taking 700 prisoners; amongst the latter, Major Szèl, who defended himself to the last, and received sixteen wounds before he could be captured."*

Battle of Moor, 30th December 1848.—"In front of Moor, about 1200 or 1500 yards from the wood, a line of heights intersects the road at right angles; on these Perczel formed his troops and placed his guns so as to command the outlets from the forest.

"He had 5000 men under his orders, consisting of four battalions of infantry, four squadrons of cavalry, and ten guns. His object was to prevent the enemy from getting out of the wood, and give Görgey time to bring up the left wing of the army (the brigade Karger) from Csàkvár, then attack the enemy at a disadvantage.

"The enemy had, at first, but one brigade up, yet they made many gallant but fruitless attempts to gain the open country, but the Hungarian artillery, playing constantly on the outlets, rendered all their attempts abortive. After two hours' fighting, however, one of the enemy's batteries established itself on a height to the right of the road, and by a well-directed fire did much execution on the Hungarian left. At this moment Ottinger arrived with his cavalry brigade (consisting of two or three regiments), dashed at the Hungarian position, and swept all before him. In an instant our bravest infantry regiments were broken and in full flight along the road to Stuhlweissenburg and Csàkvár; one of our batteries was in the hands of the enemy.

*Klaptka, *Hist. of the Magyar War.*

"Our hussars (the regiment Nikolaus), with the greatest contempt for death, threw themselves upon the enemy's horse, though six times more numerous, and saved what they could from the wreck. A desperate hand-to-hand fight ensued, but the brave sons of the Pusta* devoted their lives to stem the tide of the enemy's advance till the remaining infantry, with four guns, made good their retreat. The hussars left half their number on the field. We lost, altogether, 1400 or 1500 men killed, wounded, and prisoners, and six guns."†

If cavalry (without artillery) have been repulsed more than once in charging a square, they should not be led against the same sides of the square, or brought forward over the same ground, because the fallen horses and men form a rampart for the infantry.

Let us now consider how a cavalry officer should act to support infantry which is attacked by a numerous cavalry.

The infantry, formed in squares or columns, will rest its flanks on any available obstacle or cover which may be within reach, and such cover will further be occupied by flanking parties, who fire in on the flanks of the advancing horsemen.

If the cover on the flanks of the infantry intercept the view of the enemy's movements, they must be watched by mounted men, whilst the cavalry, formed in small columns behind, or opposite the intervals of the infantry squares, are held in readiness to act. Should artillery play on the squares, the cavalry is, as far as possible, kept out of range, or out of the line of fire. But as soon as the enemy's cavalry advance, they do the same; and when the infantry has given them a few discharges, or beaten them off, they break forth between the intervals, form, and charge, but without pursuing too far. The object is to defend the position, and they must return to their place to be ready to repeat the same manoeuvre as often as it becomes necessary.

If the object is, however, to cover a retreat, the infantry then take advantage of the repulse of the enemy's cavalry to move off; their artillery follow with the prolonge, and are protected by the cavalry.

---

*The Prairies of Hungary.

†Klaptka.

In an open plain, where there is no cover for the infantry, the cavalry forms from 300 to 400 yards in rear and on the flanks, so as to charge the enemy's cavalry in flank when they are thrown into disorder by the fire of the squares.

This is a difficult position in which to be placed when inferior to the enemy in cavalry. They will of course try to turn the flanks, and the cavalry must then take refuge in the intervals of the squares, and wait for a favourable opportunity to charge.

When retiring along a road the cavalry keeps the road, the infantry moves on both flanks; when they reach a pass or defile the cavalry moves on and takes up a position to cover the retreat of the columns.

We have not sought to conceal the difficulties that cavalry has to contend with when brought in contact with good infantry: they are often so great that the best cavalry, though well led, will fail to make an impression; still, at equal chances, the physical superiority of the cavalry ought to give them the advantage.

Nothing is more trying to infantry than a charge of cavalry; nor is anything more formidable to cavalry than an infantry fire in square. The infantry soldier knows well that if the horsemen break in nothing can save him; the dragoon again is well aware that neither riding nor manoeuvering can save him from the bullets of his antagonists.

Success on either side depends so much on the moral courage of the parties engaged, and that success has been so varied, that numberless examples might be cited to prove the case against the cavalry as well as for it.

Good infantry will generally hold its own against cavalry if these, are badly led, if their horses are tired and worn out with fatigue, if the ground is heavy and deep, and the charge cannot be made with speed and well together, or if artillery scatters the squadrons in their advance.

It would be ridiculous to infer from such examples that cavalry cannot break infantry squares: when we look into the defeats of cavalry we can often trace their want of success to circumstances of the sort just mentioned, which are generally passed over unnoticed by historians.

For our purpose let it suffice that we have shown to our cavalry that it can be done, and we leave it to them to remove all doubt on the subject so soon as they shall have an opportunity!

Appendix

The Cavalry Saddle and Bridle Proposed by the Author

The Bridle

The bridle I propose has only four buckles,—namely, two on the head-collar, one on the bit head-stall, and one on the bridoon head-stall.

The bit, bridoon, and their head-stalls, are provided with hooks and links, by which means the bits can be slipped out of the horse's mouth, for the purpose of feeding, without taking the bridle off the horse's head.

The bridoon has half-horns to prevent its being drawn through the horse's mouth, which often happens when they have the ring only.

The head-stalls are fastened to the collar by a strap and button on the brow-band, like the bridles in use with the regiment of Carabineers, and others in the service. (*Figures 4 A and B*)

The Saddle

The tree is constructed to combine the advantages of a hunting saddle with the simplicity of the Hungarian troop-saddle.

The side-boards are cut away under the man's leg, they then spread out under his seat, and are feathered and brought well off the horse's back in rear.

The hind fork is broad at the base, where it joins the side-boards, and is bevelled off to add length to the seat.

The front fork is constructed with a peak, and with points to give the tree a firm hold on the horse's back, and prevent it from turning round, as well as to bring the man's bridle-hand low.

Both forks are strengthened with iron plates.

The holes cut for the stirrup leathers leave a whole tack, to prevent the stirrup-leather from bulging or embedding itself in the pannel, and pressing on the horse's back.

Over this tree a seat of blocked leather is stretched, like that of a hunting saddle, and fastened with screws to the forks. (*Figure 5.*)

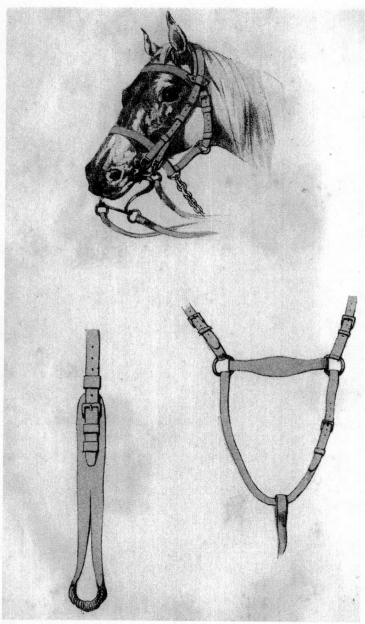

FIGURE 4 A. THE BRIDLE AND BIT.

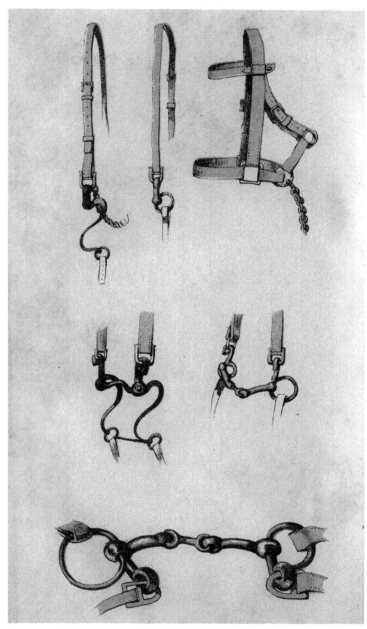

FIGURE 4 B. THE BRIDLE AND BIT.

For the Pads

Three slips of felt are slipped into a cover of serge, and put on to the sideboards with leather pockets; more of these slips of felt can be put in should the horse fall off in condition, or they can be taken out should the horse put up condition; and thus the saddle always rides even, and the tree never requires to be altered. Leather flaps are then screwed on to the tree, and the saddle is complete.

The Saddle-Cloth

Is cut to fit the horse's back, and to the outline of the saddle. It has pockets for the points to fit into, and is made of felt one inch thick, to protect the horse's back, absorb the perspiration and prevent the edges of the pads from getting hard and occasioning sore backs.

The breastplate is attached to ditto, let into the side-boards (not, as now, to the peak), and is made with a loop in front to slip in case of a fall.

The crupper is attached to a central point, to avoid the friction caused by a split crupper.

The under girth buckles on both sides, because it will wear longer, as the soldier is not obliged to girth up always in the same holes.

The girth-straps should be joined together with web girthing, to protect the seat of the saddle when put on the pegs in the stable.

The shoe-cases have loops, the strap being made a fixture to the hind-fork: the reason for this in, that the soldier in passing the strap under the hind fork is apt to leave a twist in it, which is sure to give a sore back.

The wallets are made flat, because they hold more and sit closer; they are slanted forward to give room for the man's leg.

Under the off wallet the carbine is run through a holster; it can thus be got at, and returned, in an instant. It is fastened to the peak by a strap about a yard long, which serves as a sling for the carbine when dismounted.

The valise has the troop letter and the man's number on the near side, to enable him to recognise it, and the number of the regiment on the other side, as the 10th Hussars have.

I consider a shabraque a useless encumbrance; but if worn it should made of cloth not water-proofed. Water-proofed cloth is less durable, and confines the heat to the horse's back.

Those parts which cover the valise and cloak might be lined with water-proof, but the seat should be of cloth only, and cut to fit the saddle without skirts.

Sheepskins are bad, because once wet they take many days to dry, and the heat of them is bad for man and beast.

In packing, the dragoon should be instructed to take every hard substance out of his valise. His brushes, hold-all, rubbers sponges, and forage-cap, should go in his wallets, where he can get at them at once on dismounting.

The valise should be packed so as to be quite hollow in the centre, and the centre baggage-strap shortened, in order to enable the soldier to draw it tight, and bring the kit well off the horse's back.

Boots or high-lows should be packed under the flap of the valise with the heels outwards (not inwards).

The saddle I have proposed would be much better for the men without flaps. With horsemen everything depends on their having confidence in the firmness of their seat on horseback.

Leather flaps are slippery, and do not give a firm hold to the leg; whereas, take them away altogether, substitute a double felt saddle-cloth cut square, and the men will have a stronger seat; the horses will be more under control on that account, as well as because they feel, and must obey more quickly, the pressure of the riders' legs.

The whole of the Austrian cavalry ride without flaps to their saddles; but boots or leather over-alls are then indispensable on account of the sweat from the horse.

The saddles without flaps would be lighter, and far more handy for saddling.

My attention has been directed for a long time to the constructing of a saddle for military purposes. With this end in view, I have for years past taken notes of what came under my observation, and put everything to the test of a trial when I had the opportunity.

On my return from the Russian camps last year, being ordered to join the depot troop of my regiment at Maidstone, I there constructed the first of these saddles which I now propose.*

Many hints and improvements I had from Colonel Key, 15th Hussars; for instance, the manner of joining the hind fork to the side-boards I copied from a saddle which Colonel Key obtained from an officer of Danish Hussars at

*Messrs. Gibson and Son, 6, New Coventry Street, constructed all those subsequently made, and improved much upon the original one.

FIGURE 5. THE NEW SADDLE AND TREE (*a, b, d*) AND THE OLD (*c*).

Copenhagen. The addition of points to the front fork was a suggestion of Sergeant Johnson, saddler at Maidstone, a very intelligent, clever man, with whose assistance I also contrived to improve upon the hooks and links of the bit and snaffle which I had brought from Russia.

One of the greatest improvements suggested, I believe to be the felt saddle-cloth. The blankets accumulate dirt and dust in their folds; they constantly work out from under the saddle. In camp they must always be kept ready folded, and, in case of surprise at night, they are most inconvenient, for if a man drops one he cannot refold it without assistance.

In this saddle the man can be put in a proper seat, he will have greater control over his horse, and I do think that with them sore backs in the cavalry will be of rare occurrence.

Dress of a Cavalry Soldier

The dragoon should be supplied annually with one good suit of serviceable clothes, which he should be made to wear.

At present he has too many things; the full dress is seldom worn, and never worn out; the man does all his duty in his stable-dress, whilst the kit is swelled out with unnecessary things, and the horse is over-weighted.

One pair of over-alls.

One waistcoat, say red, with sleeves and pockets.

One surcoat of blue cloth made loose, with the sleeves to unbutton to the elbow (like those in use with Spanish soldiers), and turn up when the gauntlets are put on. This, with a good great-coat or cloak, a low head-dress properly ventilated, would answer all purposes for the cavalry soldier.

Remarks on the Rank Entire System

The Rank Entire system has been again brought before the public in an article in the *United Service Gazette* of March 12th, 1852, and recommended by some of the very highest authorities, including the late Duke of Wellington, and several of our most distinguished cavalry officers. Under these circumstances it would be presumptuous in me to attempt to give an opinion of my own, but I shall adduce what I think may be considered as arguments against the rank entire system, founded on the authority of other distinguished officers of cavalry.

Seidlitz and Ziethen fixed upon *two ranks*, on *small squadrons*, on *wide intervals*.

Single-rank men have no backers sufficiently near to inspire them with confidence and perseverance; and this confidence is as necessary in the horses as in the men to induce them to rush into fire.

However good your cavalry soldiers may be, they are not *all* fit for the front rank; neither are *all* horses fit to lead, though *all* will follow.

Broken up in mêlée, the single ranks (men being equal) would be overpowered before they could get assistance, and the single-rank reserves would be again over-matched by the double-rank reserves.

A single rank, if successful, cannot spread like a double one in the pursuit, sweeping up all before them, but they will be so much scattered, that, in riding in between the retreating dragoons, their own flanks will be exposed, and the retreating horsemen, on their left, will immediately close upon them, and cut them down whilst they are endeavouring to assail the men on their right front: thus the chase will become more perilous to the pursuers than to the pursued.

A squadron of the 1st Lancers (British Legion), formed in rank entire, consisting of sixty horses, charged three hundred Carlist cavalry, pursued them three miles, and killed nearly one hundred of them. Would the same sixty Englishmen have failed to do the same had they been formed in two ranks?

To add that great essential, *rapidity*, to the movement of cavalry, keep the squadrons small, and give them plenty of elbow-room, that is, sufficient intervals.

There was no want of speed shown by the Prussian cavalry under Seidlitz.

The advantage of having an officer to command each detachment of the rank entire system can be equally shared by the two-rank system. This is simply a question of expense, and must depend upon how many officers you would attach to each one hundred men.

Those nations whose cavalry formerly acted on the rank entire system have given it up. Take the Cossacks for example.

However, whether it be in two ranks, or in one, let your cavalry be formed in small, distinct bodies, with sufficient intervals between each.

Let there be no pivot flanks, and no *right* and *left in front*, and, doubtlessly, cavalry will prove much more efficient in the field.

Proposed Organization of Cavalry

From the *United Service Gazette*, 12th March, 1853.

In the *United Service Gazette* of the 8th May last we noticed a very interesting pamphlet by Colonel Kinloch, on the subject of the Constitution of a Yeomanry Force, and especially in reference to the organization of cavalry in "rank entire." We have now the pleasure of publishing a letter from Colonel Kinloch, embracing the opinions of the late Duke of Wellington, Lord Vivian, Lord William Russell, and General Bacon, upon that momentous subject. It is peculiarly well timed, as at the present moment great interest is taken in the question of increasing the efficiency of our present weak regiments of cavalry, and of rendering the yeomanry fit for active home service.

"To the Editor of the *United Service Gazette*
"Sir,

"In a pamphlet I published last year on the Defence of the Country by means of a Volunteer Force, I ventured to recommend cavalry, and yeomanry in particular, to be formed in rank entire, instead of in two ranks, as is usual in most armies.

"I have lately found copies of the opinions of the Duke of Wellington and several cavalry officers on this subject, which I could not lay my hand on at the time I wrote the pamphlet. These opinions were addressed to General Anthony Bacon (who commanded the cavalry of Don Pedro, in Portugal, in 1833-34), and who was good enough to give me copies of them."

I.

From F. M. The Duke of Wellington, K.G.
"Strathfieldsaye, 20th November, 1833.

"Cavalry is essentially an offensive arm, whose use depends upon its *activity*, combined with its steadiness and good order.

"I think that the second rank of cavalry, at the usual distance of close order, does *not increase* the ACTIVITY of the cavalry. The rear rank of the cavalry does not strengthen the front rank, as the centre and rear ranks do the front rank of the infantry. The rear rank of the cavalry can augment the activity or even the means of attack of the front rank only by a *movement of disorder*.

"If the front rank should fail, and it should be necessary to retire, the second or rear rank is *too close to he able to sustain the attack* or to restore order. The second rank must be involved in the defeat and confusion, and the whole must depend upon some *other body*, whether of cavalry or infantry, to receive and protect the fugitives.

"I have already said that the rear rank can only augment the means of the first rank by a movement of disorder.

"This is peculiarly the case if the attack should be successful. In all these cases the second rank, at a distance sufficiently great to avoid being involved in the confusion of the attack of the front rank, whether successful or otherwise, could aid in the attack, or, if necessary, cover the retreat of the attacking party, and thus augment the steadiness and good order of the cavalry as a body; while, by the absence of all impediments from the closeness of the rear rank, the activity of the front rank would be increased.

"It cannot be denied that, till required for the actual attack, the less cavalry is exposed the better. My notion of the distance of the lines of cavalry was as much as a cavalry horse could gallop in a minute; the second line should pull up at a walk when the first charges; the third and other lines in columns should deploy, or be used according to circumstances.

"I conceive that the one-rank system would require a change, not only in the discipline, but in the organization of the cavalry. If I am not mistaken, it would render the use of cavalry in an army *much more general than it is at present*.

(Signed) Wellington"

II.

Extract from a letter to General Bacon, from Lieut. General Sir Hussey Vivian, G.C.B. (afterwards Lord Vivian, and Master-General of the Ordnance).

"I rejoice to find Don Pedro's cavalry has fallen under the orders of a man so capable of leading them. I again congratulate you on your very high and honourable station. In short, I feel confident you will do credit to yourself, your cavalry, and to your Peninsular education.

"I quite entirely agree with you in all you say of the value of the lance.

"As to the rank entire system, I am by no means certain that it would not always be a good thing, if on advancing to an attack, or standing in line, the rear ranks were to form a reserve at the distance, say of eighty or one hundred yards; when so circumstanced they would be much better able to follow up an advantage gained by, or to repel a successful attack of, the enemy on the first rank. The fact is, *that the second rank is but of little use but to fall over the first*.

"Let me congratulate you on your brilliant success and gallant conduct at Leyria. I will run over and pay you a visit.

(Signed) R. Hussey Vivian"

III.

Extract from a letter to General Bacon, from Lord William Russell, Colonel commanding 8th Hussars.

"I wish you joy of your promotion and command. The cavalry will, I have no doubt, be in excellent order in your hands; but don't be rash—they are too numerous for you to cope with, and their horses are better fed.

"Anything that proves the efficiency of the single-rank system is interesting to me; and it certainly was thoroughly proved on the 16th October (1833), when your force of cavalry imposed on more than treble your numbers; this quite destroys the argument that a single rank 'looks so weak,' and 'invites the enemy to charge.' Your adversaries were not to be tempted on the 16th. Keep notes of all the occurrences; we will one day put them in print.

"I am delighted to find that Vivian (Sir Hussey) looked with a more favourable eye on the system. Depend upon it they will all come round. He wants to get off on the *mezzo termine* of leaving the rear rank behind. This I entirely disapprove, because the *rear rank so left would have no one to command it, and cavalry depends entirely on its officers.*

"There is no doubt that, if cavalry is to act in one rank, a different organization is necessary. You must turn your mind to this, as the end of the war brings to your aid the practical reflections you can make now. The Duke of Wellington is in our favour, but the prejudices of the cavalry officers are difficult to be overcome.

"I send you up —————, and if you can convert him you will do wonders. He never in his life gave up a once formed opinion.

"Try Head's (Sir Francis) plan with the lasso for your guns.

(Signed) William Russell"

IV.

Extract from a letter from General Bacon
"11th September, 1835.

"My Dear Kinloch,

"I hope you stick to it 'rank entire.' Depend upon it, it is the most efficient way of using cavalry. You are quicker and have more reserves. Enclosed are some extracts* respecting my system with cavalry. Lord Anglesea, Brotherton, and many other officers I could name, agree with me. As I am the only man who has tried it, I give you a few of my reasons.

"In one rank all movements are made with greater precision and more rapidity than in two.

"When cavalry has to re-form after a charge it is effected more readily and far quicker, for each man gets at once to his own troop, and, if such formation be required under fire, the value of *quickness* will admit of no argument against it.

"I have tried this in presence of a superior enemy very frequently, and at times when hotly pressed, and under a heavy fire of artillery and musketry.

"A charge in one rank will be more rapid, consequently more likely to suc-

*The above from the Duke of Wellington, etc.

ceed, than one in two ranks, because the horses are more at liberty, not likely to be cramped by the endeavours of the rear rank to get to the front, and the men will have a more free use of their arms; every one will do his duty; *skulkers* cannot so easily pull up, and such are found in all armies.

"In advancing in line for any distance (and before an enemy you have rarely a fine open country) the intervals are never preserved between squadrons, and it frequently happens that a line of two ranks towards the centre becomes a disordered column; in re-forming, a rear rank is never sure of its 'telling off.'

"In all columns I should form my second rank in a column in rear of my first, that is, as a second regiment, and this will always be easy by keeping, when in line, a distance equal to the depth of a close column; you may always close your lines if you think it desirable, and, when about to form columns, it is only to open your ranks, or, instead of a column of squadrons, to form on the centre a contiguous close column of half-squadrons.

"Another great advantage in the system is, that all your ranks are commanded by officers. Whenever you are asked for a squadron, remember it is a troop, and if you send two troops they are two squadrons, and they become a proper command for a major.

"I could give you many other reasons for the system, but I hope to be with you, and perfect that which I have begun, and with a fair portion of success.—Believe me, etc.,

(Signed) A. Bacon."

A squadron of the 1st Lancers of the British Legion in Spain, under the command of Major Hograve, and consisting of only sixty horses, charged three hundred of the Carlist cavalry (after they had defeated several squadrons of Christino cavalry), pursued them above three miles, and killed nearly one hundred of them. The squadron of the 1st Lancers was formed in "rank entire," the Carlists in two ranks, and thus proved triumphantly the efficiency of that system.

From the above opinions of distinguished and experienced officers, supported as they are by facts on the few occasions the "rank entire" formation has been tried, it appears worthy of consideration whether it should not now be acted upon in the British army.

I find that, in the *Regulations for the Drill and Exercise of the Yeomanry Cavalry*, they are recommended to adopt the "rank entire" system, which I was not aware of when I wrote my pamphlet last year. This formation is particularly suitable for yeomanry and all irregular and half-disciplined cavalry, on account of its great simplicity and freedom of action. As the Duke of Wellington "conceived that it would render the use of cavalry in an army much more general than it is at present;" as Sir Hussey Vivian said "that the second rank is but of little use but to fall over the first;" and as General Bacon, observes "a troop becomes a squadron, and each rank is commanded by officers;" it appears that the effective strength of our cavalry may be greatly increased, *if not fully* DOUBLED, by adopting the rank entire system. The principal change in the organization called for by this alteration appears to be a *small* and *economical* addition to the officers.

Captains commanding troops will, when in line, command squadrons, for a troop will occupy the same front in single rank that a squadron does in two. (The interior economy of the troop remains as before.) The officer who commands two troops or squadrons when in line will have a fit command for a field officer.

I would therefore suggest, that, in order to carry out the advantages of the "rank entire" system, and render it thoroughly efficient, a second major should be restored to each cavalry regiment; and as a half-troop will become a half-squadron, there should be another subaltern to each troop.

Perhaps the two or three senior captains should have a higher rank than merely captain, as they may sometimes command two troops or squadrons; the higher rank of chef d'escadron, as in other armies, or brevet major, might be given to them.

The only additional expense, therefore, would be a second major for each regiment, and a second lieutenant to each troop.

A very small expense certainly, if, as it is assumed, the cavalry will be greatly increased in efficiency.

Let the troops be increased to fifty horses per troop, with the above addition to the officers, and adopt the "rank entire" formation; then our cavalry will be ready for any service that might be required of them; but at present the regiments are very weak in horses, and in double ranks are only about two good squadrons strong.

I must apologise for troubling you so long; but as the economical increase of the army is the great question of the day, and cavalry appears not much noticed, I trust the foregoing remarks and opinions on the organization and efficiency of that chivalrous and important arm of the service will not he considered uninteresting to your military readers.

I have the honour to be your obedient humble servant,

John Kinloch
Logie, 7th March, 1853.

P.S.—The following sentences are extracted from my pamphlet, before referred to.—J. K.

"It was on his (General Bacon's) recommendation that I adopted this, which appeared to me to be the proper and common-sense formation of cavalry (and raw cavalry in particular), in the 1st or 'Reyna Ysabel' regiment of lancers in the British Legion in Spain, which I had the honour to organize and command in 1835-36; and to that simple formation I attribute the very creditable and efficient manner in which they could go through the manoeuvres of a field-day, and do their duty in the field, after very little practice and drilling. After being broken in a charge, or dispersed in pursuit, a squadron in rank entire may 'rally' and 'tell off' in much less time than the front and rear ranks would take to scramble into their places; and thus much time, so valuable in cavalry movements, would be gained.

"Many smart soldiers dislike being in the rear ranks, and feel themselves thrown in the background; they are apt to become careless, and merely to follow their front-rank files, without knowing or caring what is going on; whereas in rank entire every man is under the eye of his officers, and MUST be on the *qui vive* and wide awake. Every man has an equal share in the attack, which is not the case with two ranks. Rank entire may *appear* loose, and show more 'daylight' between the files, but is not in *reality* more loose; on the contrary, cavalry accustomed to work in rank entire will be found to be better closed together than with two, though the two ranks help to 'fill up' better, and make them appear closer and more solid.

"I consider that a charge of cavalry in rank entire, on fair ground, fit for a good gallop, will 'hit harder,' and every man and horse 'tell' with greater effect, than if the rear rank were treading on their heels. The rear rank cannot give much assistance to the front, but they may actually incommode them. In the rank entire system, young and untrained horses will be quieter and steadier, and not so likely to be lamed by the rear rank treading on the front rank horses' heels, and they, in turn, 'lashing out,' and laming those behind. It may be said that the two ranks have answered very well on former occasions, but it may be a fair question to ask, whether half the number would not have done as well? or whether the same number in two lines, instead of two ranks, would not have done better?"

Troop-Horses and Officers' Chargers

Before I left India, some very interesting trials were made at Madras, by order of the Commander-in-Chief, General Sir George Berkeley, the object of which was to test the capabilities of the troop-horses, as well as the relative merits of entire horses and geldings for the purposes of war.

Three trials were made.

The first with two regiments of Native Regular Cavalry, one of stallions, one of geldings.

The next with two troops of Horse Artillery.

The third, and last, with two hundred English dragoons (15th Hussars); one hundred riding stallions, and one hundred mounted on geldings. This squadron marched upwards of eight hundred miles—namely, from Bangalore to Hyderabad, where they remained a short time to take part in the field-days, pageants, etc. They then returned to Bangalore, four hundred miles, by forced marches: only one rest-day was allowed them, and the last six marches in were made at the rate of thirty miles a day. They brought in but one led horse; stallions and geldings did their work equally well, and were in equally good condition on their return. The question was, however, decided in favour of the latter, because they had been cut without reference to age, and only six months before the trial took place.

The English cavalry in India is well mounted. On an emergency any one of these Indian regiments would gallop fifty miles in a pursuit, leave few horses behind, and suffer but little from the effects of such exertion. The horses on which they are mounted are small but powerful. The Arab, the Persian, the Turcoman, the horses from the banks of the Araxes, are all unrivalled as war-horses. I have seen a Persian horse fourteen hands three inches carrying a man of our regiment of gigantic proportions, and weighing in marching order twenty-two and a half stone: I have seen this horse on the march above alluded to, of eight hundred miles, carrying this enormous weight with ease, and keeping his condition well; at the crossing of the Kistna, a broad, rapid, and dangerous river, the owner of this horse (Private Herne, of C troop) refused to lead the animal into the ferry-boat to cross but, saying "An hussar and his horse should never part company," he took the water in complete marching order, and the gallant little horse nobly stemmed the tide, and landed his rider safely on the opposite bank.

An officer in India made a bet that he would himself ride his charger (an Arab, little more than fourteen hands high) four hundred measured miles in five consecutive days, and he won the match; the horse performed his task with ease, and did not even throw out a wind-gall. The owner, an officer of the Madras Artillery, died shortly afterwards.

General Daumas relates that the horses of the Sahara will travel during five or six days from seventy-five to ninety miles a day, and that in twenty-four hours they will go over from one hundred and fifty to one hundred and eighty miles, and this over a stony desert. Diseases of the feet and broken wind are almost unknown amongst them.

What would become of an English cavalry regiment if suddenly required to make a few forced marches, or to keep up a pursuit for a few hundred miles!! Their want of power to carry the weight, and want of breeding, makes thorn tire after trotting a few miles on the line of march.

Our cavalry horses are feeble; they measure high, but they do so from length of limb, which is weakness, not power. The blood they require is not that of our weedy race-horse (an animal more akin to the greyhound, and bred for speed alone), but it is the blood of the Arab and Persian, to give them that compact form and wiry limb in which they are wanting.

The fine Irish troop-horses, formerly so sought for, are not now to be procured in the market. Instead of the long, low deep-chested, short-backed, strong-loincd horse of former days, you find nothing now but long-legged, straight-shouldered animals, prone to disease from the time they are foaled, and whose legs grease after a common field-day.* These animals form the staple of our remount horses.

Decked out in showy trappings, their riders decorated with feathers and plumes, they look well to the superficial observer; but the English cavalry are not what they should be. If brought fresh into the field of battle, the speed of the horses, and the pluck of the men, would doubtless achieve great things for the moment; but they could not *endure*, they could not follow up, they could not *come again*.

All other reforms in our cavalry will be useless unless this important point be looked to. It is building a house on the sand to organise cavalry without good horses. Government alone could work the necessary reform by importing stallions and mares of eastern blood, for the purpose of breeding troop horses and chargers for the cavalry of England.

It is said that a government stud is opposed to the principle of competition. What competition can there be amongst breeders for the price of a troop-horse

On the Condition of our Saddle-Horses; T. Hatchard, Piccadilly.

when by breeding cart-horses they obtain forty pounds for them when two years old? How could they possibly afford to rear animals with the necessary qualifications for a cavalry horse of the first class? To breed such horses a cross must first be obtained with our race-horses: this would entail a large outlay of capital; and when the good troop-horse was produced, the breeder could not obtain his price for him.

The rules of our Turf encourage speed only, and that for short distances. Horses are bred to meet these requirements, and from these weeds do our horses of the present day inherit their long legs, straight shoulders, weak constitutions, and want of all those qualities for which the English, horse of former days was so justly renowned.*

I had heard of fine horses in Russia, but I complacently said to myself, "Whatever they are, they cannot be as good as the English." However, I went to Russia—and seeing is believing. Their horse-artillery and cavalry are far better mounted than ours; and their horses are immeasurably superior in those qualities which constitute the true war-horse—namely, courage, constitutional vigour, strength of limb, and great power of endurance under fatigue and privation.

The excellent example set by Sir George Berkeley in India might be followed up at home with great advantage to the service; the capabilities of our cavalry horses of the present day should be severely tested, and the saddles should be tried and experiments made to ascertain how sore backs may be avoided.

On the Condition of our Saddle-Horses.

FURTHER READING

Jon Coulston

THE FOLLOWING DISCUSSION OF recommended literature should be read in conjunction with this volume's introduction. It is intended to allow the reader to follow up issues of interest concerning Nolan and his writings, the place of cavalry on the battlefields of the 'long' nineteenth century, and the British Army in the Crimean War which both tested Nolan's theories and ended his life.

There is just one definitive study of Nolan's career and his importance for the British Army, H. Moyse-Bartlett, *Nolan of Balaklava and his influence on the British Cavalry*, Leo Cooper, London, 1971. Alongside Nolan's *Cavalry* his *Training the British Cavalry* (Napoleonic Archive (www.napoleonic-archive.com), n.d., reprint of 1852 edition) should be consulted. His campaign journal survives in the collection of the National Army Museum (Chelsea, London), as does the cloak which he lent to William Howard Russell. One might also compare two other nineteenth century treatises: Col. Ardant du Picq, *Battle Studies. Ancient and Modern Battle*, English edition, Macmillan and Company, New York, 1921 (du Picq was killed at the head of his regiment in August 1870); Captain Loir, *Cavalry. Technical Operations. Cavalry in an Army. Cavalry in Battle*, H.M.S.O., London, 1916 (employing examples from 1870). The ancient Greek treatise by Xenophon, *On the Art of Horsemanship* (Xenophon, VII, *Scripta Minora*, Loeb Classical Library, Harvard University Press, Cambridge and London, 1968; early 4th century BC) is the ur-text for works on horse riding and cavalry, and formed the basis for some of Nolan's views on horse treatment. The Byzantine genre of cavalry treatises with diagrams may be further pursued in Ilkka Syvänne, *The Age of the Hippotoxotai. Art of War in Roman Military Revival and Disaster (491-636)*, Acta Universitatis Tamperensis 994, Tampere, 2004.

There are several very accessible general modern studies of 19th century warfare. Geoffrey Best's *War and Society in Revolutionary Europe. 1770-1870* (Fontana History of War and European Society, Fontana, London, 1982) is a good introductory text. M. Glover, *Warfare from Waterloo to Mons*, Book Club Associates, London, 1980, and B. Holden Reid, *The Civil War and the Wars of the Nineteenth Century*, Cassell, London, 1999 are well-illustrated run-throughs of the wars and issues. P. Howes, *The Catalytic Wars. A Study of the Development of Warfare, 1860-1870*, Minerva Press, London and Atlanta, 1998, despite its restricted title, actually reviews developments from the Napoleonic period onwards, with special reference to the American Civil War and the wars of German unification.

There is a need for overarching literature dealing with nineteenth century cavalry, but developments may be followed through a number of publications. A historical treatment of British cavalry is provided by the Marquess of Anglesey's magisterial series, *A History of the British Cavalry 1816-1919*. Volumes 1 and 2 cover the period from the Napoleonic Wars, through and beyond the Crimean War describing establishments, reforms and campaign histories (Volume 1, *1816-1850*, Leo Cooper, London, 1973; Volume 2, *1851-1871*, Leo Cooper, London, 1975). The *History* goes on to deal with the First World War actions in Palestine (Volume 5, *1914-1919. Egypt, Palestine and Syria*, Leo Cooper, London, 1994). The examination of Napoleonic British cavalry by Ian Fletcher is an antidote to Wellington's attitudes which does much to restore the reputation of the 'mere brainless gallopers' (*Galloping at Everything. The British Cavalry in the Peninsular War and at Waterloo, 1808-15. A Reappraisal*, Spellmount, Staplehurst, 1999). In general see P.J. Haythornthwaite, *Napoleonic Cavalry*, Sterling Publishing Company, New York, 2001. British interest in the lance was maintained through the published findings of trials: R. Hervey de Montmorency, *Exercises and Manœuvres of the Lance*, Naval and Military Press with the Royal Armouries, Uckfield, n.d. (reprint of 1820 edition).

For the performance of Austrian and Hungarian cavalry in the 1848 Revolutions see Istvan Deak, *The Lawful Revolution. Louis Kossuth and the Hungarians, 1848-1849*, Phoenix Press, London, 2001. Later continental cavalry deployments are examined in excellent studies by Geoffrey Wawro (*The Austro-Prussian War. Austria's War with Prussia and Italy in 1866*, Cambridge University Press, Cambridge, 1996; *The Franco-Prussian War. The German Conquest of France in 1870-71*, Cambridge University Press, Cambridge, 2003).

The American experience, which proved so crucial for the development of cavalry fire-power and dismounted tactics, is represented in a vast amount of literature on the Civil War. In general see Brent Nosworthy, *The Bloody Crucible of Courage. Fighting Methods and Combat Experience of the Civil War*, Carroll and Graf Publishers, New York, 2003. Specific works on cavalry tactics include S.Z. Starr, *The Union Cavalry in the Civil War*, Leicester University Press, Leicester, 1971; E.G. Longacre, *The Cavalry at Gettysburg. A Tactical Study of Mounted Operations during the Civil War's Pivotal Campaign, 9 June–14 July 1863*, University of Nebraska Press, Lincoln and London, 1986; D. Evans, *Sherman's Horsemen. Union Cavalry Operations in the Atlanta Campaign*, Indiana University Press, Bloomington and Indianapolis, 1996; James Pickett Jones, *Yankee Blitzkrieg. Wilson's Raid through Alabama and Georgia*, University Press of Kentucky, Lexington, 2000; E.J. Wittenberg, *The Union Cavalry Comes of Age. Hartwood Church to Brandy Station, 1863*, Brassey's Inc., Washington, 2003. Rush's Lancers are discussed by Starr (above) and by E. Wittenberg (*Rush's Lancers. The Sixth Pennsylvania Cavalry in the Civil War*, Westholme, Yardley, 2006). A meticulous study of firearm technology is provided by J.G. Bilby, *A Revolution in Arms. A History of the First Repeating Rifles*, Westholme, Yardley, 2006. A detailed analysis of U.S.–Mexican War weaponry is provided by C.M. Haecker and J.G. Mauck (*On the Prairie of Palo Alto. Historical Archaeology of the U.S.-Mexican War Battlefield*, Texas A&M University Military History Series 55, Texas A&M University Press, College Station, 1997) alongside a clear picture of hesitant Mexican cavalry leadership and tactics at the Battle of Palo Alto. Overwhelming American mounted firepower is further demonstrated in a series of small actions of the 1830s-1840s in Texas, as described by Bill Groneman (*Battlefields of Texas*, Republic of Texas Press, Plano, 1998) and Robert M. Utley (*Lone Star Justice. The First Century of the Texas Rangers*, Berkley Books, New York, 2002).

A manual based on the British experience of colonial wars, recommending the most effective ways in which mounted troops could be deployed, is C.E. Callwell, *Small Wars. Their Principles and Practice*, H.M.S.O., London, 1906 (University of Nebraska Press, Lincoln and London, 1996). It makes reference to the American Plains Wars, for which see further Robert M. Utley, *Frontier Regulars. The United States Army and the Indian, 1866-1891*, University of Nebraska Press, Lincoln and London, 1984; Bill Yenne, *Indian Wars. The Campaign for the American West*, Westholme, Yardley, 2006.

For Frederick Burnaby see his travel accounts: *On Horseback Through Asia Minor*, Oxford University Press, Oxford, 1996; *A Ride to Khiva. Travels and Adventures in Central Asia*, Oxford University Press, Oxford, 1997. The literature concerning the Victorian British Army is extensive, but attention may be drawn to three works which impart some of the flavour and atmosphere of military service: Byron Farwell, *Mr Kipling's Army*, W.W. Norton and Company, New York and London, 1981; John Strawson, *Beggars in Red. The British Army, 1798-1889*, Pen and Sword Books Ltd, Barnsley, 2003; I.F.W. Beckett, *The Victorians at War*, Hambledon and London, London and New York, 2003. More specialised works concentrating on technological developments, the need for reform, and the impact of the Crimean War on the army are provided in J. Sweetman, *War and Administration. The Significance of the Crimean War for the British Army*, Scottish Academic Press, Edinburgh, 1984; H. Strachan, *Wellington's Legacy. The Reform of the British Army, 1830-54*, Manchester University Press, Manchester and Dover NH, 1984; H. Strachan, *From Waterloo to Balaklava. Tactics, Technology, and the British Army, 1815-1854*, Cambridge University Press, Cambridge and New York, 1985 (outstanding contribution); H. Strachan, 'The British Army and 'modern' war: the experience of the Peninsula and of the Crimea', in J.A. Lynn (ed.), *Instruments, Ideas and Institutions of Warfare, 1445-1871*, University of Illinois Press, Urbana and Chicago, 1990, 211-37. For British army reviews and ceremonial events see Scott Hughes Myerly, *British Military Spectacle from the Napoleonic Wars through the Crimea*, Harvard University Press, Cambridge, Mass., 1996.

The Crimean War in general has received detailed attention from historians up to the present, but Kinglake's official history, first published in 1868, is still a mine of information: A.W. Kinglake, *The Invasion of the Crimea. Its origin, and an account of its progress down to the death of Lord Raglan*, Vols 1-9, William Blackwood and Sons, Edinburgh and London, 1901 (now in course of being republished in the The Michigan Historical Reprint Series, University Library, University of Michigan). To this may be added Russell's contribution as the leading in-theatre newspaper reporter (William Howard Russell, *Russell's Despatches from the Crimea, 1854-1856*, edited with an introduction by N. Bentley, Andre Deutsch, London, 1966; R.J. Wilkinson-Latham, *From our Special Correspondent.Victorian War Correspondents and Their Campaigns*, Hodder and Stoughton, London, 1979).

The tempo of publication quickened towards the anniversary year of 2004, some of the more recent contributions including: P. Kerr, G. Pye, T. Cherfas, M. Gold and M. Mulvihill, *The Crimean War*, Boxtree-MacMillan, London, 1997; W. Baumgart, *The Crimean War, 1853-56*, Arnold, London (Oxford University Press, New York), 1999; R.B. Edgerton, *Death or Glory. The Legacy of the Crimean War*, Westview Press, Boulder and London, 1999; T. Royle, *Crimea. The Great Crimean War, 1854-1856*, Little, Brown and Company, London, 1999; I. Fletcher and N. Ishchenko, *The Crimean War. A Clash of Empires*, Spellmount Limited, Staplehurst, 2004; A. Massie, *The National Army Museum Book of the Crimean War. The Untold Stories*, Sidgwick and Jackson with The National Army Museum, London, 2004; C. Ponting, *The Crimean War. The Truth Behind the Myth*, Chatto and Windus, London, 2004; J. Spilsbury, *The Thin Red Line. An Eyewitness History of the Crimean War*, Weidenfeld and Nicholson, London, 2005; D.S. Richards, *Conflict in the Crimea. British Redcoats on Russian Soil*, Pen and Sword Books Ltd, Barnsley, 2006.

Special mention must be made of the catalogue of the 2004 National Army Museum (Chelsea, London) Crimean War exhibition (A. Massie, *A Most Desperate Undertaking. The British Army in the Crimea, 1854-56*, National Army Museum, London, 2003). This is an invaluable collection of essays which discuss the war, its battles, the army, public reactions and the aftermath, with a sumptuously illustrated catalogue of the exhibition's documents, artworks and artifacts, including the actual written order carried by Nolan from Airey to Lucan.

In this connection one Crimean episode, the Charge of the Light Brigade, has been minutely examined in a number of books: C. Woodham-Smith, *The Reason Why. The Story of the Fatal Charge of the Light Brigade*, Penguin Books, London, 1958; M. Adkin, *The Charge. The Real Reason Why the Light Brigade Was Lost*, Pimlico, London, 1996; T. Brighton, *Hell Riders. The True Story of the Charge of the Light Brigade*, Henry Holt and Company, New York, 2004. Two penetrating biographies have researched the colourful and controversial leader of the charge: Saul David, *The Homicidal Earl. The Life of Lord Cardigan*, Little, Brown and Company, London, 1997; Donald Thomas, *Cardigan. A Life of Lord Cardigan of Balaklava*, Cassell and Co, London, 2002.

British uniforms and equipment of the Crimean War have been lavishly published, partly in connection with film-making (see below). See W.Y. Carman (introduction), *Dress Regulations 1846. The Uniform of the British Army at the beginning of the Crimean War*, Arms and Armour Press, London, 1971; R. Wilkinson-Latham, *Crimean Uniforms 2. British Artillery*, Historical Research Unit, London, 1973; *Uniforms and Weapons of the Crimean War*, B.T. Batsford Ltd, London, 1977; M. Barthorp, *Crimean Uniforms. British Infantry*, Historical Research Unit, London, 1974; J. Mollo and B. Mollo, *Into the Valley of Death. The British Cavalry Division at Balaklava, 1854*, Windrow and Greene, London, 1991; D. Featherstone, *Weapons and Equipment of the Victorian Soldier*, Arms and Armour Press, London, 1996. For the swords of the period the definitive study is B. Robson, *Swords of the British Army. Regulation Patterns, 1788-1914*, National Army Museum, London, 1996, supplemented by C. Martyn, *The British Cavalry Sword from 1600*, Pen and Sword Books Ltd, Barnsley, 2004. Such research has been greatly facilitated by the surviving photographic archives from the Crimean campaign: H. Gernsheim and A. Gernsheim, *Roger Fenton. Photographs of the Crimean War*, Arno Press, New York, 1973; P. Hodgson, *Early War Photographs*, Osprey Publishing Ltd, Reading, 1974; L. James, *Crimea, 1854-56. The War with Russia from Contemporary Photographs*, Hayes Kennedy, Thame, 1981.

Lastly, mention must be made of two fictionalised treatments of the Battle of Balaklava. During the course of his Crimean adventures, the anti-hero of George MacDonald Fraser's Flashman novels stands with the Thin Red Line and charges with both the Heavy and Light Brigades at Balaklava (*Flashman at the Charge*, Barrie and Jenkins Ltd, London, 1973). Raglan, Lucan, Cardigan, Campbell, Scarlett, Russell and Nolan all appear prominently. Of course 'Flashy' survives the Valley of Death to be captured by the Russians.

The film *The Charge of the Light Brigade* (UK, 1968) was not originally a box-office success, but since release has become recognised as a classic of British cinema and acquired a cult following. Contemporary with the Viet Nam War, it is an anti-war and anti-class parody drawing directly on Victorian newspaper coverage and the critical public perception of the army in the Crimea, employing both animation and live action. Nolan is a central character with much of his writings being directly quoted in dialogue. He is played dashingly, but perhaps too charismatically, by the late David Hemmings (*Blow Up*, *Barbarella*, *Gladiator*, *Gangs of New York*, *League of Extraordinary Gentlemen* etc.). Many detailed liberties were taken with history. Raglan (John Gielgud) is exaggeratedly bumbling, and Cardigan (Trevor Howard) is made to look both ridiculous and stupid. Three years of meticulous research went into accurate visual recreation of the period, but at the last moment the director, Tony Richardson, took the ludicrous decision to dress the entire Light Brigade in the cherry-red overalls worn famously and solely by Cardigan's 11th Hussars (his 'Cherubims/Cherry-bums'). The researchers were so enraged and humiliated by this blatant anachronism that they determined to publish a series of uniform studies based on their work. Thus were born the invaluable books by Barthorp, Mollo and Mollo, and Wilkinson-Latham cited above. This is a level of historical commitment only equalled in the last forty years by work on *Master and Commander: the Far Side of the World* (US, 2003).